SCIENCE ISN'T EVERYTHING -

MEMOIRS OF A SCIENTIST

LLOYD KEMP

Copyright 2009, Lloyd Kemp
Permission is freely given
to quote from this work,
provided due acknowledgement is made.

ISBN: 978-1-905795-51-2
Published by Aspect Design,
and printed and bound at their premises
89, Newtown Road, Malvern, WR14 1PD

An Appreciation

Though these Memoirs were begun several years ago, the bulk of them has been written during the last twelve months, something which, at my age, I would never have achieved without the help and encouragement of others. In particular there have been five, whose contribution has been crucial. Having said that, I need to add straightaway that in each case the contribution, from one aspect or another, has been unique. For that reason if for no other, the order in which I record the nature of their contributions is no indication at all of their relative importance. Indeed, it is true to say that in their own way they have been *equally* important, and it would be invidious even to imply that it was otherwise!

Thus, bearing in mind the importance of my poetry in the latter half of the Memoirs, I would like to put on record that it was Dr Linda McHugh, my GP, who, way back in 2004, when I was near the rock bottom of a deep depression brought on by the sudden loss of all musicality in my hearing, said to me that she thought that 'the last 18 months was worth suffering, to effect the changeover [from music] to poetry.' My comment in my Journal was that 'She could hardly have put it more strongly, knowing full well, as she did, what I was going through.' [From that very first poem of 'the new era' – 'Mother and Child' – she had requested that I emailed each new poem to her, 'hot off the press'.] 'I left the surgery head high again, As a GP, she certainly knew all about suffering in its multitudinous forms, and she had shared my own special, and private agonies.' And, in all this, she was aided by my Shiatsu practitioner Mercedes Nuñez, who, likewise, 'from that very first poem of "the new era" ', gave me crucial support, which I needed so badly, at that critical time.

Then there was Bruce Hedland-Thomas, my long-standing friend in Perth, Australia, together with my friend of many years and former colleague Dr Barry Barber and his partner Maureen, who, without any collusion between them, and whenever my enthusiasm for my task showed signs of waning, would give me a nudge (and sometimes a prod!) from the direction of Malvern in Worcestershire and of Perth in Australia (a sort of 'pincers movement' I called it!) to motivate me again.

But there was more to it than that. Barry and Maureen introduced me to the digital age in commercial printing: what might be regarded as an extension of desktop publishing, enabling small printing runs to be commercially viable. The author simply produces a master .pdf file of his book, from which, with no further ado, small to modest-sized editions can be run off. As for Bruce, he 'walked with me all the way'. Right from the start, I have sent him each new chapter as it was written (or re-written!), and usually within a matter hours I would receive his emailed comments and suggestions, often with a list of typographical errors for good measure. Over some six years in all, he carried out the role of an editor, whose comments were thoughtful, very much to the point, and always encouraging, whenever I showed signs of wilting before the magnitude of the task I had set myself. What friends they all were, and, of course, still are!

And finally, there was my son John's contribution as a climax to the whole project. His Adobe Photoshop skills amount almost to wizardry! From a very mixed bag of photographs and documents, some of them dating back 70 or 80 years, he spent a lot of time optimizing their appearance and size, to provide a total of 20 Plates illustrating the text. What a difference they make! He has also played an important part in producing the master .pdf file for submission to the printer.

All five, who have contributed so much - and, indeed, made sure that these Memoirs would finally see the light of day - are already very well aware of my appreciation of their help and encouragement, but I need to take this opportunity of acknowledging it publicly.

Lloyd Kemp, Bath, November 2009

Foreword

According to my dictionary, 'Memoirs' are 'a written account of one's memory of certain events or people'. An 'Autobiography', on the other hand, is said to be 'an account of a person's life written by that person.' Therein lies the reason for deeming this piece of writing to be 'memoirs' rather than an autobiography: I could work up little enthusiasm for the latter, with its necessity to leave hardly a stone unturned, and, moreover, to turn them over in some sort of chronological order. More importantly, I do not regard what I have done simply as an account of my life. That, for me, would be rather boring, and too preoccupied with self. No! – let it be regular strolls down Memory Lane, to come upon side-turning after side-turning waiting to be explored – with a surprise round almost every corner; and yes - for me as well as the reader.

Throughout my life, I have been driven by 'the creative urge', and whilst writing these Memoirs, I have realized all over again that being creative *does* 'keep you going', both physically and mentally. And it's not simply a matter of the quality or significance of what I've created in science, painting, poetry, or writing: it's the act of creativity itself. I had long felt this to be so, when, quite fortuitously, I lighted upon a TV programme reporting a substantial research into this very matter. When you get to my age, and are seen to be still pretty active both mentally and physically, folk are inclined to ask, 'What's the secret?' To which my reply is, 'It's no secret – I suppose it's just doing something creative with hands, head, or heart. It doesn't have to be ground-breaking, mind-boggling or whatever - it just needs to be *you* being *yourself*, creatively.'

I think it was the character Mona Lott, in Tommy Handley's famous wartime comedy series ITMA, whose catchphrase was, 'It's being so cheerful uz keeps me going.' Were I such a character, for sure *my* catch-phrase would have to be, 'It's the need to be creative that keeps *me* going.'

Lloyd Kemp, Bath, November, 2009

Chapter 1

It *has* to be 1916, as I was born in 1914, and it is my earliest memory.

At the time, we were living in a cul-de-sac of around a couple of dozen houses. To be strictly accurate, it was a cul-de-sac only so far as vehicles were concerned, for there was a short footpath at the end, that enabled pedestrians to exit the cul-de-sac without retracing their steps. To a mere two-year-old, the footpath seemed to offer access to a whole new world beyond it, and one day it beckoned irresistibly. Somehow, I managed to escape from the house, and toddled to the end of the road and along the footpath to that big, wide world beyond. I had only just emerged from it when a voice, which seemed to come from heaven, hailed me.

'And where do you think *you're* going?' As a two-year-old, I was in no position to argue with God. Instead, I looked up, to find that the voice had actually come from Tom Phipps, our milkman, perched high up in his milk cart. He came to our door daily, when I would watch him as he lifted the lid of a big metal can containing the milk, and measured out a pint or half a pint using one or other of two ladles, stowed in the can, together with the milk. As I look back, I realize with some horror that in transit the milk would have slopped around the handles of the ladles, handles which would have been in the milkman's hands only moments before, hands which had been involved in receiving money, and giving back the change. Nevertheless, before the howls of disapproval break out, we need to remind ourselves that there is a school of thought that would argue that nowadays we may well be living in an environment *too* sanitized for the good of our immune systems...

To continue: even if I had had a plausible destination, I was given no chance to reveal it. 'Just you go back 'ome as fast as you can, or I'll –' I didn't stay long enough to hear what my punishment would have been. I took to my heels, and ran as fast as my legs would carry me back down the footpath again. And as I ran, I wetted myself – which is probably why the memory is so vivid!

Despite that, other early memories were much more frightening. The little cul-de-sac was called Grove Road, and it was near the centre of Stevenage, about 25 miles north of London, in those days a little country town of about 5000 inhabitants. I was tucked up in bed and asleep when my father grabbed me out of bed, wrapped me in a shawl, and took me out onto the landing just outside my bedroom door, where there was a south-facing window. 'Look! Look!' he cried, pointing to something in the night sky. 'It's a zeppelin. And it's on fire. And - and there are people in it.' The word zeppelin meant nothing to me, but the fact that it was on fire, with people in it, certainly did, and the horror of it is with me still, the better part of a century later.

There was another memory of the Great War, which embedded itself in my mind, and has been with me ever since. Once again I had been put to bed, but this time I was lying awake in the dark, when there was a strange whistling sound, which seemed to pass right over the house, followed by two enormous bangs that rattled the bedroom window violently. The momentary silence which followed was broken by an equally unfamiliar sound – of my father whistling as he walked along the path beside the house, leading to the back door. Apparently, he told my mother afterwards that he had been trying to create the impression that nothing unusual had happened, despite those bangs that had frightened the life out of me. They had been caused by bombs, of course – the first and only ones to be dropped near the little town, which certainly harboured nothing by way of a military target, save a battery of searchlights on the outskirts. My mother, who came running up to my bedroom, referred to the bombs as 'things' which some nasty men had dropped by mistake, and my young mind tried to imagine what you could drop which would make such a noise, but I found no answer to the riddle.

The next day the whole family went for a walk – that is, my mother and father, with myself toddling along with my sister Vera, four years older than me, and my baby brother, only weeks old, in his pram. The purpose of the walk turned out to be nothing less than the inspection of two large holes which had appeared as if by magic in a field not that far away. I was all ears, and gathered from what was being said that the holes had been caused by the things the nasty men had dropped, but the nature of such objects which, just by dropping them, could make holes of that size baffled me. It turned out from what was being said that the men in the zeppelin were nasty men too, and deserved what had happened to them.

The burning zeppelin – horrendous though it was as an image of death, was remote and somewhat unreal. But there was worse to come: in a matter of weeks, and just as winter set in, my baby brother developed bronchitis, and, in a matter of days, was dead. My father, who was a skilled cabinet maker who had failed his medical and had been reduced to making ammunition boxes in the local factory, made the coffin, bringing to bear on it all his frustrated skills. There were no hardwoods available, so he made it out of deal, and painted it white, after which my mother had lovingly padded it with silk which she had by her.

After that, they had put the coffin – without the lid on – in an armchair in the sitting room, with my baby brother's body at rest in it. They told me he had gone away, and might not be coming back, thinking to break the news gently to me. As the sitting room door had no lock on it, they apparently told me not to go into it for the present. Despite that, my curiosity had got the better of me, and a day or two later, when my father was at work, and my mother and sister were in the garden hanging out the washing, I sneaked in. The curtains were tightly drawn (as was the custom in those days, when there had been a death), adding to the air of mystery, and creating a sudden fear in me. And there, in the open coffin, I could dimly discern the body of my baby brother. He hadn't gone away at all, I thought, with rapidly growing horror, as I viewed the body, its lifeless state shockingly reminiscent of that of a mouse dumped on the hearthrug by our cat.

The worst of all, though, the very worst, had yet to come.

Consequent upon my baby brother's death, my mother had a breakdown. Years later she told me that she had 'lost all her emotions', but at the time it seemed to me that she had gone into her shell, just like one of the snails I played with in the garden, and seemingly as dead. In fact she told me that the winter had turned to Spring before, like the plants in the garden, she at last showed signs of coming back to life again.

The effect on me had been insidious. Having, to all intents and purposes, got my mother back from the dead, I was taking no more chances, and young as I was, and in my childish way, I assumed personal responsibility for her happiness and well-being, feeling it incumbent on me to do all within my power to prevent such a thing from ever happening again. And this included accompanying her on her weekly visits to my baby brother's grave.

For reasons buried in history, the little church was well beyond the outskirts of the town, and I enjoyed walking there, between the rolling cornfields. I enjoyed the views too, from the churchyard, when we got there. But it was the ritual itself that, like the war, came to burden my spirit with feelings far beyond my years.

It began with a visit to the water trough, tinged always with anxiety, lest there was no watering can in the little cupboard alongside it; and for the rest of my life I have remembered the hiss of the ball valve as the tank refilled, after my mother had dipped the can into it. As she knelt by the grave, she would give me the dead flowers to put in the waste basket a few graves away, whilst she carefully arranged the new bunch in the terra cotta pot, which she afterwards filled with water from the can. The rest of the water she would sprinkle over the grave for the benefit of the grass, though I was convinced that it was to fulfill some esoteric need of my dead brother.

Week after week, month on month, year upon year, the ritual continued unchanged throughout my childhood, until - like the wording on the gravestones all around - every detail had become carved on my memory, woven into the very fabric of my young life.

I have said little of my father so far, and there is no doubt that my mother and her problems dominated my childhood. It is memories of my father's skills as a cabinet-maker, however, which come flooding back. I loved to escape down the garden path to his workshop on a Saturday afternoon, when my mother had gone out shopping. I would shuffle my way through the wood shavings, often ankle deep on the floor, like the carpet of leaves in the woods in the autumn, but which, unlike the leaves, smelled, not of decay and death, but of the sap of living wood. I would perch myself on a stool at the far end of the workshop, and watch my father as he drove a jack plane the full length of a piece of timber, held tightly in the jaws of a bench vice. The smell of the resinous wood was a heady one that I loved, and the crisp curl of the shavings as they left the throat of the plane to join their companions on the floor fascinated me.

'Look at that!' he would say, pointing to a subtle and beautiful pattern that the plane had revealed in the grain of the wood. I would jump down from my stool then, and go over to him, and he would indicate the pattern with a gesture of the back of his hand, which seemed almost as loving as my mother, when she stroked my head. He indeed had the true love of the craftsman for his materials.

We were a poor family, and not only had my father made my baby brother's coffin, but he had also made a substitute for the usual marble surround to his grave, consisting of four simple wooden rails, mounted in beautifully-carved corner posts, and, of course, needing to be painted regularly. Even as I write, I can still see so plainly, in my mind's eye, every detail of that quasi white marble surround, which my father cared for as lovingly as my mother cared for the grave itself. My parents' families have long since died out, as have all connections with Stevenage itself – my last visit in the 1960s, when I attended my Aunt Cis's funeral. And, inevitably, I find myself wondering about the fate of that surround, constructed, and cared for, with such love. Without a doubt, it will have rotted away, leaving no trace, the patch of ground probably not even identifiable any longer, as a grave.

Why, I asked myself, are these memories of my baby brother's grave so vivid still, nearly ninety years on? The reason was not difficult to find. You see, whenever another coat of paint was needed, I was always eager to go with my father – it being a far less fraught occasion, compared with the weekly visits with my mother to the graveyard. I suppose the comparison points up a difference between my parents that was of general significance within the family: my mother was an ardent chapel-goer of the Primitive Methodist ilk, whilst my father was not given to formal observances, chapel or otherwise. Yet, in his unassuming way, he was in essence a deeply spiritual man: each time his plane revealed unusual beauty in the grain of a piece of wood a revelation, and his joy in it almost akin to worship.

Another difference between visiting the graveyard with him rather than my mother was that I got to ride on a saddle fitted to the crossbar of his bicycle! – a mode of transport for her precious son that my mother had only recently agreed to, and soon regretted. She would watch with not a little apprehension as the two of us sped down the short hill to the end of the road. I was unable to wave goodbye as, excitedly, I clutched the handlebars with both hands. It was still a novelty for me, as I sat astride the saddle between my father's arms. 'Bye, Mum,' I would shout, as she stood watching us as we disappeared round the corner at the bottom of the road. I am sure that each time she watched as we set off, she regretted all over again having ever agreed to such a mode of transport for her precious son.

My final memory of those earliest years comprises a vivid, visual one - of both my mother and father in bed for days, victims of the influenza pandemic which swept the World at the end of the war, and of a neighbour bringing in food for us. I have no recollection of being in bed myself, so I suppose I must have escaped the infection that killed some 20,000,000, worldwide.

Chapter 2

My mother and my father both came from large families, my mother having seven brothers and sisters, and my father five – but how different the two families were! This was borne in upon me from a very early age. The aunts and uncles on my father's side were earthy folk, always capable of calling a spade a spade and something else as well if necessary, whilst on my mother's side a severe and, indeed, somewhat puritan atmosphere prevailed. There was more to it than that, however: my recollection of my father's family is of healthy and robust people living for the moment, whilst a certain otherworldliness prevailed in my mother's family. One reason for this was that most of them were devout Primitive Methodists, but another and more immediate reason was the prevalence of chronic ill health in the family. By the time I arrived on the scene, my uncle Will had long since been bedridden, severely crippled by rheumatoid arthritis. And I can just remember my grandmother Lavinia, before she, too, retired to bed for the rest of her days, with a supposed cancer of the throat. In spite of that she lived on, untreated, for a number of years. Then there was Aunt Annie, who wore leg irons, presumably another victim of arthritis; Aunt Lizzie, too, who, even to one of my tender years, seemed the odd one out – overweight to the point of ungainliness, dribbling from the corner of her mouth as she talked, her accent and vocabulary betraying her mental backwardness. I sensed that she was an embarrassment to the family, although, now, she would almost certainly have been diagnosed as myxoedemic. Finally, there was Aunt Alice, who suffered from breast cancer, and underwent a mastectomy.

Why should I remember such detail, in the case of my Aunt Alice? Well - I was accompanying my mother on what was virtually a nightly visit to the family home, to sit for a while, in turn, with her bedridden mother, and brother Will. Alice was visiting from London, and after a brief exchange of words with my mother she prepared to show her what the mastectomy looked like, compared with the healthy breast. There were others present, including Annie and brother Arthur, who immediately protested that it was not seemly for me to see this (note - it was of the *healthy* breast that he spoke). Alice replied, reassuringly, that she didn't mind at all, adding that I was too young to be interested in women's breasts. No-one, however, gave any consideration to the effect that seeing a brutal scar (it must have been getting on for a foot long) - might have on me: I can still see it so vividly in my mind's eye, nearly eighty years on.

Incidentally, Alice was something of a black sheep as far as the rest of the family was concerned. She had, it seems, 'gone into service' in the household of a London 'gentleman', who had got her pregnant. It was thus that I came to have an illegitimate cousin named Cecil, who retained his mother's maiden name of 'Young', although, years later, the father did marry Alice (after his wife had died).

Apart from Will, my mother had three other brothers, but before turning to them I must mention an extraordinary enterprise that Will pursued successfully, and on a worldwide basis, over many years, despite his bedridden state. As I have said, the whole family was of Primitive Methodist persuasion, and Will certainly made a notable and very unusual contribution to the air of otherworldliness that pervaded the whole household. He wrote religious poems – strictly speaking hymns, since they were the requisite few verses long, and written in the most common metres used in hymns. There were dozens and dozens of them, and they kept coming! So where did the enterprise come in? It was quite simple really, and certainly ingenious. He would have dozens of copies of any particular 'hymn', complete with an appropriate title, printed on a tinted sheet of roughly A6 size, and each particular pile of copies was kept in its own compartment in a shallow chest subdivided into dozens of compartments (no doubt made for the purpose by brother Arthur, the carpenter – see below). This provided ready access to them, and they looked rather pretty in their multiplicity of pastel shades.

Those three other brothers of my mother's were George,

Frank, and, as I said, Arthur. George was very well spoken, and, indeed, well-educated (largely by attending evening classes), and given to quoting Latin tags in the middle of an otherwise mundane conversation. Frank lived in London, where he had a wife and family. Consequently, I saw far less of him, and learned little about him, except that he was a carpenter. Arthur was, too, but I saw much more of him, since he continued to live in the family home, even after he married Cecilia ('Cis'), who came from Birmingham, and I can still remember how her accent stood out among the local (Hertfordshire) accents of the rest of the family.

We are talking about the early 1920's, and Arthur must have been one of a very small band of enthusiasts who dabbled in radio in those very early days. To begin with he built 'crystal' sets, every detail of which I can still see in my mind's eye: the little glass tube containing the crystal and the so-called 'cat's whisker'. There was also a cardboard tube on which was wound many turns of fine black-enamelled wire, along which slid a knob by means of which the number of turns in use could be increased or decreased, for tuning purposes. The only available signal was from '2LO', and Arthur, wearing headphones, could nightly be seen bent over this contraption, busy finding a 'good spot' on the crystal by means of the cat's whisker, and tuning in to 2LO by sliding the knob along the turns of wire. Should anyone dare to speak during these operations, he would emit a loud 'Shoosh,' glaring fiercely at the offender. He later graduated to a 'valve' set and a loudspeaker, and his equanimity was restored.

Despite the differences between the two families, they did have one thing in common – strangely, they each owned a 'corner shop'. On my father's side the shop was run by his sister Ruth, and on my mother's side by Annie. Even though the two shops were each only about five minutes' walk away from us, my mother chose to deal almost exclusively with Annie: it was a rare event for me to be sent round to Ruth's shop for something – my mother never went herself.

From an early age I sensed her animosity towards my father's family. Indeed, she made no secret of it in the way she spoke of them, and inevitably some of it rubbed off on me. I suspect that it was all to do with the inability of Primitive Methodism to agree with my father's description of a spade, so to say.

With hindsight, I think my sister Vera favoured my father's uncomplicated approach to life, with the result that often – *too* often – there was a feeling that our family was split down the middle, with my sister taking my father's side, and I, from a very early age, rushing to my mother's defence. That probably sounds somewhat melodramatic, but it is a sad fact that though they were united in the practical aspects of bringing up their children, and in that respect were good parents, so far as the emotional background was concerned, they were often sadly lacking. There were frequent angry and loud-voiced quarrels, the cause of which to this day I have failed to understand, and for that reason they were particularly distressing.

Well do I remember one occasion - a dismal, late afternoon in winter, when I was in my early teens. The quarrel was threatening to get out of hand, so much so that all I could think to do by way of protest was to go to the piano in the sitting room, where I thumped out, as loudly as I could, the first few bars of Beethoven's Fifth Symphony: di di di dah, di di di dah. It worked, and the raised voices subsided. What made these quarrels particularly distressing was that, basically, the marriage was a stable one, something which I must have sensed, and in the context of which the quarrels were incongruous - indeed, inexplicable. However, it was a great consolation to me that as my parents reached middle age, the quarrels became few and far between, and eventually ceased altogether. Furthermore, as they became elderly, the two families themselves became reconciled, and I remember how gratified I was when my parents, who had long since moved away from Stevenage, told me they had invited Aunt Ruth to stay with them – something hardly conceivable in earlier days.

Of course, there were the ramifications of the two families, namely the crop of cousins they produced. There was, for instance, Cecil, already mentioned. He enlisted in the army at a very young age early in the war, and became a stretcher-bearer. As often as he could – and it would seem almost daily – he wrote what one can but refer to as a 'dispatch' to Uncle Will, with lurid details of all that being a stretcher-bearer involved. Will, despite being bedridden, still managed to write - with the aid of a small wooden board to support the paper - and he lovingly, and with great care, copied each of Cecil's letters into a large notebook, of which, in the end, there were several, which he finally had bound into one large volume. This came into my possession, following Cecil's death, and not wishing such a document to languish on my bookshelves, I donated it to The Imperial War Museum North, in Manchester. I'm sure that it would provide researchers with just the kind of detail they sought.

Cecil survived the war unscathed, despite the dangers to which he was exposed as a stretcher-bearer. Although, strictly speaking, he was my cousin, he was some nineteen years older than I was, and I always referred to him as 'Uncle Cecil'. His position was, indeed, unique within the family: all my other uncles being some twenty years older than Cecil, or more.

Under government sponsorship he became a student at Bristol University, where he took a Teaching Diploma, and it was during that time that he met Frances, a florist (she had come from Natal), who became his wife. She was very pretty, and extremely vivacious, and the two of them together could hardly have provided a greater contrast, compared with all my other aunts and uncles!

Their first visit to our home was made memorable for me by a wooden castle they presented to me (which I can still see, in my mind's eye), complete with 'lead' soldiers. Thereafter, they became regular visitors, especially at Christmastime, when I literally sat at Cecil's feet whilst he regaled us with tales of his exploits as a stretcher-bearer. He was an excellent raconteur, and also a good pianist, and we had wonderful sing-songs round the piano, with the aid of a 'Daily Express Community Song Book', which is still in my possession. And the games we played! – from Musical Chairs to Murder, and that hilarious game called Consequences, in which entries were made on a strip of paper, and then hidden, entry by entry, by folding the paper before passing it on. The end point was reached when the papers were unfolded, and read out. There was nothing puritan or Primitive Methodist about Cecil and Frances. How well do I remember the gales of laughter with which Frances greeted such statements as 'Mr So-and-So (probably living in our road) met Miss So-and-So (ditto) in Lover's Lane at midnight, and he said to her '- - -', and she said to him '- - -', and the consequences were... etc, etc. ...' I suspect that this particular game was very much frowned on by my mother's family (and very much enjoyed by my father's!).

It is worth pursuing the 'Cecil and Frances saga' a little further, as they had an important influence on me in other ways. For one thing it was clear to me from very early on that they believed that life was there to be enjoyed, providing a welcome corrective to the impression I was gaining, particularly from contact with my mother's family, that life was a grim business, to be endured rather than enjoyed. And, in 1925 I won a scholarship to the local Grammar School ('Alleyne's'), some twenty years after Cecil had done so; and sharing the same *alma mater* created a special bond between us.

Alas, as the years passed, both Cecil and Frances lost their high spirits, and became involved in a whole sequence of 'way-out' movements, from spiritualism to Jehovah's Witnesses and the British Israelite cause, to say nothing of an off-shoot of Egyptology, which claimed to be able to forecast the future from measurements made on, and inside, the pyramids. What did I learn from all that? – simply the dangers of extremism in any form: the two of them changed so dramatically that it would hardly be an exaggeration to say that they were no longer the same people.

Of the other cousins, there was one that I shall certainly never forget. Strictly, he was the husband of Lizzie's daughter 'little Lizzie' - a cousin by marriage, you might say. His name was Reg, and, like Cecil, he had fought in the war, but, alas, without Cecil's good fortune: he had fallen victim to sleeping sickness, following service in Africa. He was a frequent visitor to our home, and would often trail off into a state of deep drowsiness, even in the middle of a sentence. It was another of those things to add to the list - of an uncle crippled almost literally into knots by arthritis, and an aunt who was forced to wear leg irons to get about at all, to say nothing of Aunt Alice's amputated breast, and Lizzie's myxoedema, a formidable list indeed, against which Cecil's and Frances's fun and laughter did their best to compete. Even so, was it the sombreness of that list that made them finally take refuge in the corridors of the pyramids, or the false security of fundamentalism, such as that of the Jehovah's Witnesses; or must we look elsewhere?

Perhaps. After twenty years of marriage, they had produced no offspring. Then, at the late age of 39, Frances had become pregnant, only to suffer a miscarriage. For both of them it was a terrible disappointment, and one which might well have accounted for the change that came over them. We shall never know.

Chapter 3

I was nearly six when I started school – I think my mother was reluctant to part with me! I attended an Infant School until I was eight. Strangely, I remember little or nothing of those years. What I do remember so well is what might almost be called the 'rite of passage' which ended them. The class that was to move on to the 'elementary school' was lined up to form a 'crocodile', and then marched off. The elementary school was the better part of a mile away: a few hundred yards along a beautiful avenue, followed by a right turn onto a footpath leading past allotments to a little lane. A short distance along the lane, and then a left turn onto another footpath, leading to Walkern Road. After crossing this, the final leg of the journey was yet another footpath, and leading, at long last, to our destination, the 'elementary' school.

If one's intention had been to choose a route between the two schools which would produce a maximum sense of leaving the old school far behind, to start again in an entirely new world, I think the route I have described might well have been the outcome. Indeed, the sense of leaving behind one world, to enter an entirely different one, is with me still, after all these years.

One unusual subject on the curriculum at the elementary school was gardening – and attached to the school was an area divided into allotments, each allotment having a small group of boys alocated to it. We cultivated both flowers and vegetables. In those days allotments were a normal part of most working men's lives, and played an important part in keeping the family well fed. Apart from gardening, I remember nothing unusual about the curriculum at the elementary school, or the way it was taught, but I do remember that the so-called 'Scholarship Examination' was very much its target. This was taken at the age of eleven, and the handful of successful pupils moved on yet again - to Alleyne's Grammar School, less than a stone's throw away from the Infant School where it all began. Full

Plate 1 Classroom in 1558 block, Alleyne's Grammar School

circle, you might say!

The Grammar School was founded in 1558 - indeed, as the School Song has it, 'E'er Bess to the throne ascended'. The founder was The Rev Thomas Alleyne, who, at the same time, founded schools at Stone and Uttoxeter. Again, as the School Song has it:

> 'Stevenage, Stone, and Uxeter, too,
> Schools at them all will found, sir,
> But Stevenage was ever the home of my youth,
> I'll be buried in Stevenage ground, sir.'

During the seven years that I attended the school, the original (1558) building was known as the 'Science Block' - consisting of a laboratory and a single classroom, Plate 1 being a photograph of the latter. Note the gaslights! – and the bas-relief reproductions at the back. As can be imagined, their presence dominated the whole classroom, though I have to admit that I have long since forgotten what they depicted, but the central one may well have been - appropriately enough - Socrates teaching. Since writing those words, I contacted The Old Boys' Association, who confirmed that its formal title is indeed 'Socrates Teaching the People in the Agora', the original bas relief belonging to Manchester University. Incidentally, he seemed to be delighted to inform me that to the best of his knowledge I was their 'oldest old boy'!

Relatively modern buildings had been added, which included a block of four classrooms, two on the ground floor, and two on the first floor, each pair of classrooms capable of being opened up into a single space by sliding back a dividing partition. It was in the ground floor space thus created that the whole school (consisting of a mere 120 or so pupils!) assembled each morning, after gathering in serried ranks in the open space under the gymnasium, one side of which consisted of three open brick arches. By modern standards the school was ridiculously small, its 120 pupils divided into six forms. There is no doubt, however, that the small classes were a great advantage from the teaching point of view. Indeed, the Sixth

Form Science and Arts groups consisted of a mere five or six pupils apiece, approximating to the individual teaching associated with university tutorial systems. The majority of the pupils were fee-paying, the sons of local business men and farmers, but each year a handful of pupils from local 'elementary' schools were admitted, after passing the 'Scholarship' exam, some of these schools being in nearby towns and villages.

Alleyne's is situated alongside what used to be known as the Great North Road - just as it leaves the town. How well I remember singing, so heartily, the relevant verse in the School Song! -

> Years roll onward, and Youth has fled,
> And you're fighting Life's battle for fame, lad –
> Think of the school by the King's highway,
> That taught you to play the game, lad.'

As we sang it, in our middle teens, it was almost impossible to imagine a time when youth had fled - yet, for me, as I write, that happened more than half a century ago!

The School Song was a rollicking affair, the words written by the then headmaster, Mr Thorne, and set to music by the music master, Mr Hazzard, and even recalling such fragments as I have been able to, it has certainly stirred the emotions again. But, there, School Songs are surely meant to be like that – almost tear-jerkingly so.

A school as small as Alleyne's is bound to create and nurture a real community spirit, denied to the modern 'comprehensive' school, twenty times as big. By the same token, one's recollections of one's schooldays are likely to be much more personal. In particular, for me, having spent my Sixth Form days in the Science group, it is natural that some of my most vivid memories will be of one Frank Slow, the science master.

For example, he was renowned for his puns. Seeing an apple core left on a bench in the laboratory, he was heard to say, 'Whoever left that there, kindly remove it. We don't want any cores for complaint lying around here.' And there were plenty more like that...

Mind you, he didn't have it all his own way. In those days there was a make of motorcycle called the 'Ivy', which Mr Slow sported, as his means of getting to school. One day, in a General Science class involving the botanical species of the same name, he commented that 'ivy creeps slowly along the ground, and needs frequent support.' He was *not* amused by the titter of laughter that went round the class!

Then there was Miss Wilkie, who taught English to the lower forms. She had been a governess to the late Queen Mother, and spoke impeccable English, and could be quite scathing at times, when we country lads fell short of her standards. In those days there was no such thing as a fountain pen - *or* a biro! The desks were fitted with so-called 'inkwells', let into the desk top. Each form had its 'ink monitor', whose task it was to keep the inkwells topped up. On the day in question, he had failed in his duties, and one of the class, finding his inkwell bereft of ink, duly put up his hand. 'Well?' said Miss Wilkie, impatiently. 'Please, miss, may I have s'mink.' 'S'mink, boy - what's *s'mink?*' It must be good teaching, if I have remembered this little exchange, after all these years...

Mr Thorne, the Headmaster, had near-black hair, with a moustache to match, and, to small (and larger!) boys, his appearance and bearing was little short of terrifying. His study was a room in his own house, the door of which opened onto the playground, hard by the school gates. It was part of the ritual of morning assembly for him to sweep out of his study, carrying a large black bible, and to stride up the playground, his gown trailing behind him in the slipstream he was creating. The door into that double classroom, where the whole school awaited his arrival almost apprehensively, was in charge of a boy known as the door monitor. His unenviable task was to keep a sharp look-out through the window, and to open the door at exactly the right moment, so that the headmaster could enter and continue unchecked down the aisle between the rows

Plate 2a Alleyne's Football Team 1931 - 1932

Plate 2b Sharing the Victor Ludorum Cup

of desks to his own desk, perched on a sort of wooden dais.

Now, a small number of boys lived in the neighbouring village of Knebworth, and - beyond that - in Welwyn Garden City, and they travelled to school by train (on the London and North Eastern line, as it was called then). There had been some cases of carriage upholstery slashing (yes! – even in those days), and Mr Thorne had had a visit from the railway police as part of their enquiries, and he had assured them that none of his boys could *possibly* be responsible for such a thing.

One morning, however, he came storming, rather than striding, up the playground, to the consternation of the door monitor, who only just managed to get the door open in time. Down the aisle he swept, and no sooner had he taken up his position behind his desk when he thumped the Bible down onto it with a loud thud. With no further ado, and without attempting to open the Bible, and white with anger, he thundered out, 'It is a *farce* holding prayers in a school like this.' (The railway police had found the culprits, and Mr Thorne had been proved sadly wrong in the assurances he had given them.)

Although, basically, I was what in those days would have been called a 'bookworm', especially if the book was about any aspect of science, I was, nevertheless, very keen on sport and gymnastics. 'Sport', at Alleyne's, meant football, cricket, and athletics. In those days, the positions occupied on the field by the eleven members of a football team were easily recognizable, and immutable – five forwards, three half backs, two backs and a goalkeeper, and, being fairly fleet of foot, I soon found myself allotted the position of 'outside right', which I retained throughout my days at Alleyne's. Plate 2a shows the School Team for 1931/1932, my last year at the school.

As for cricket, I regarded myself as a batsman rather than a bowler. When fielding, I chose wicket-keeping – that is, until I failed to stop a ball from a fast bowler with my hands but with my front teeth instead. I'm glad to say that in due course they all tightened up again, but I had quite lost my appetite for wicket-keeping... I can't move on from cricket at Alleyne's without recording an incident which comes back to me as a vivid image, almost as detailed as a photograph. It is of my mother standing just inside the front door of our house, with myself outside on the path, having turned to face her. It was games afternoon, and I stood there, wearing my school cap and blazer, and long white flannel trousers. It was the first time that I'd worn 'long 'uns', and I felt so conspicuous in them that I was refusing to go any further. My mother remonstrated long and hard with me, and I did move off eventually, feeling that everyone was looking at me, as I wended my way up the High Street to the school.

As regards athletics, my fleetness of foot made me a sprinter, a pretty fast one at that, with the result that one year, near the end of my school days, I proudly shared the Victor Ludorum cup with one of my form mates (Plate 2b).

As far as gymnastics are concerned, we did nothing to compare with what little teenage girls get up to these days: just vaulting over the 'box', and performing, upside down, on the 'wall bars'. The gymnasium, with the open assembly area underneath it, was a relatively recent addition, butting onto the end of a mid-nineteenth-century two-storey building which had comprised a library at ground level, and a dormitory on the first floor. There had long since ceased to be any boarders, and the dormitory had become a changing room, with a flight of wooden steps – some eight or nine of them - leading up to the gym. The changing room had a trussed roof, the beam of one of the trusses being above the lower part of the stairs, and a little above head level as one stood on the top step.

It was something we dared each other to do - namely to jump from the top step towards the beam, wrapping one's fingers over the top of it, and swinging from it, before finally letting go, to land on the floor adjacent to the bottom step.

One morning I duly leapt from the top step, to the cheers of my form mates, cheers which were suddenly silenced as my hands hit the beam, but my fingers failed to wrap round it. My body continued under its own momentum, and was almost horizontal as I started to fall. The first contact I made was with the end of my spine, as it hit the edge of the bottom step. Instantly my whole body was smothered with 'pins and needles'.

'I've broken my back! I've broken my back!' I yelled. As is the wont of boys, the laughter broke out again. Of, course, I ought to have gone to hospital for an X-ray, but I sat it out for the rest of the day at school, the pins and needles gradually subsiding, and I said nothing at all to my parents when I got home - so conditioned was I by my mother's extreme hypochondria. In fact, I spoke of the fall to no-one, no-one at all, for over twenty years, when, at last, I allowed a doctor friend to X-ray me, fearful though I was, of what he might find. To my great relief, he found nothing except some slight displacement in the pelvic area. Why was I still so concerned? Simply because my mother had drilled into me as a young child that *a severe blow to the body could cause cancer...*

Perhaps this is where I should say a little more about my mother's fears and phobias. Apart from a general hypochondria, she suffered from a severe phobia about cancer. I could hardly have been ten years old when she began asking me to feel the odd lump on her neck or wherever – and I would reassure her (which I sensed she desperately needed to be), before I had any idea what cancer was!

(In fact, in the earliest photograph of myself in my possession - Plate 3 - I must have been close to that age; and as I contemplate it, I cannot help feeling the unfairness of such young shoulders being asked to bear that kind of responsibility. But sadly, such unfairness is not uncommon in the lives of young children.)

To resume: a few years on, and my knowledge of magnetism and electricity was such that I was able to construct a so-called 'shocking coil', popular in those days, which comprised a high-ratio transformer (every turn wound by hand!) with a vibrating contact in the primary circuit, which produced relatively large voltage pulses in the secondary. To each end of the secondary winding was attached a flexible wire, to the other end of which was soldered a copper tube. With a copper tube

Plate 3 The Author, about 10 years of age

grasped in each hand, one was subjected to the high-voltage pulses (but very little current), which caused muscles to contract and relax with each pulse. Moreover, if we each held just one of the copper tubes, the circuit could be completed by my putting my free hand on one of my mother's lumps, in the hope that it would disappear. It didn't, of course, and was almost certainly associated with arthritis. It hardly needs to be said that I, too, had problems with cancer, long before I even knew what it was. Little did I know that that early conditioning would eventually determine my life's work.

An unusual item on the curriculum at Alleyne's was carpentry, the paraphernalia for which was housed in a hut separate from the other school buildings. Mind you, we never actually *made* anything! We simply learned how to construct the various joints – such as the tenon and dovetail - used to fasten two pieces of wood together. The catch was that before you were allowed to tackle the joint itself, you had to plane the two pieces until they were identical 'rectangular parallelepipeds'. Well do I remember one of the boys in my class – Tressider by name – who ought never to have been allowed to wield a jack plane. Why? - because, by the time he had managed to create two rectangular parallelepipeds (if he ever managed to), one, or both of them had become too thin for use in constructing a joint! Poor Tressider! He tried so hard, and invariably burst into tears when the verdict was delivered.

I took the carpentry classes for granted, especially as my father was a cabinet maker, but when I think about it now, as I write, it does seem a bit odd: learning to harness horses to a plough might have been more appropriate.

One of the more vivid memories of my school days I owe to the fact that Alleyne's, in some way or another, came under the wing of Trinity College, Cambridge. Thus it was that one year I received as a 'Maths and Science' Prize a copy of A. N. Whitehead's 'Science and the Modern World' (then only recently published) from the hands of Sir J. J. Thomson, who, thirty years or so before, had discovered the electron - the first sub-atomic particle to be identified. How proud I was – to shake hands with a man who had become something of a legend by the time he came to give me my prize. He was a frail, white-haired old man, who created another of those photographic memories for me, as I see him in my mind's eye delivering his address – and, as far as one could tell, with eyes closed for most of the time! It was a unique occasion for me, evidenced by the fact that I have no recollection at all of *any* of the other dignitaries who officiated at a Prize-giving.

    I cannot end this account of my days at Alleyne's without recording the fact that, whilst there, I formed a friendship with one Clifford Walker, who came from the neighbouring village of Knebworth, a friendship which came to an end only a few years ago, when Clifford died. Needless to say, Clifford will be popping into, and out of the rest of my story.

Chapter 4

I left Alleyne's in 1932, and it says something about the modest targets the school set itself, when I say that a classmate (Norman Blow) and I were the first to go up to university for some decades. Norman went up to Cambridge, whilst I sought, and obtained, a place at King's College, London. The reason for our going our separate ways was a simple one: my parents couldn't afford supporting me at Cambridge. As it was, the School gave me a Bursary of £50 a year. To present-day University students' ears, that may sound a ludicrously small sum of money. However, it's my guess that it is equivalent to several thousands of pounds in these days. (Don't forget: £4 a week was a good living wage then.)

Well do I remember being interviewed by one E V Appleton, Professor of Physics. He was a somewhat rotund, rather short, ruddy-complexioned man, informal in speech and manner, looking more like a successful farmer than a university professor. He had a sense of humour, too. He took one look at my Application Form, and then fixed his eyes on me across his desk, and, with a broad grin on his face, said, 'Your father's a good Liberal, isn't he?' The comment was prompted by the fact that I had four Christian names – Lloyd Asquith Winston Ewart, which had always been a great embarrassment to me. The manner in which I had come by them was little consolation, either. Apparently, it was my father who proposed them, and my mother, appalled at the idea, dared him to do so, and, according to my mother, he promptly went out and did just that. However, that afternoon, when I faced the Professor across his desk, I believe they may have stood me in good stead, judging by the enthusiastic way he had referred to them: he, too, was probably a good Liberal.

One of the advantages of going to a London College was the fact that I could live at home and commute daily to the College, which was far less expensive than living in, at Cambridge. Having said that, the *dis*-advantages are evident:

there was very little opportunity of joining in the social life of the College, which has to be regarded as the loss of part of the overall education provided by the College. I would make it more general than that, by saying that College life and home life don't mix very well. I can give a practical, though light-hearted example of that. One's room in College is a space – albeit a small one - over which one exercises virtually complete control. For example, if one wishes to work on into the small hours, one can do so in splendid isolation simply by putting a notice on the outside of the door, such as 'Working, don't disturb.' Not so for me, at home. If I needed to work late, my dear mother would insist on staying up with me – 'For company', she would say. She would sit in a nearby armchair, at least knowing better than to talk. Instead, as the night wore on, she would invariably fall victim to yawning her head off, and she would soon have me doing it as well – an entirely counter-productive outcome to her offer of company!

Living at home, and commuting each day to College, also meant that my experience of College life in general was severely limited for the first three years. (I did have a postgraduate year in residence, but more of that a little later.) There were the usual College 'Rag Days', which, with the College situated in the Strand in the heart of London, could cause problems for the authorities, and I have to place on record that it was on one of these Rag Days that I had my one and only experience of having the seat of my trousers and the collar of my jacket seized by a policeman bent on 'moving me on'. I was not a keen participant on such occasions, which included such childish and potentially dangerous things as throwing fireworks at the horses of mounted policemen. There must be some good psychological explanation for that kind of regression to childish behaviour, but it still wouldn't justify it. Fireworks were regularly used in an equally frightening manner at Student Union Meetings, when firecrackers would be thrown across a densely packed hall. Another dangerous tradition was practised on College Sports days, when hundreds of students travelled to the

College Sports Ground in the backs of open lorries. The game was to collect tram-car destination boards, by flipping them from their slots as we passed the tram. With the boards quite thick, and at least 2 metres long, an accident could easily have been caused if the board failed to land in the lorry: those that did, going a long way towards covering the Students' Common Room walls, as time went on. Hey ho! - what carefree days they were!

I cannot leave the subject of student pranks without describing what was known as 'table-lifting', although this was something which I didn't experience until that fourth year in residence at 'The Platanes', the student hostel in South London. Dinner was served on polished oak tables some 12 feet long and several feet wide, with places laid with a full complement of cutlery, water glasses and jugs, etc. for maybe six students down each side, and one at either end. Suddenly, and without warning, two or three students would begin to lift the table in their vicinity. In a matter of a mere second or so, everyone else at the table would have to follow suit, to prevent cutlery, glasses, water jugs, and the like, landing in their laps. But the real skill came when it had to be slowly lowered, without mishap, to the floor again. Needless to say, table-lifting incurred a fairly hefty fine, but it was deemed to be worth it.

Staying with that postgraduate year in residence, another recollection of student life is of the regular game of billiards in the basement of the hostel, after studies were over for the day. (If snooker existed in those days, it certainly didn't enjoy the popularity it does today – I don't think I'd even heard of it then!) A strange, strongly tactile memory I have of those games: I haven't chalked the tip of a billiard cue since then – but how well I remember the rather 'squeaky' feel it had. It used to set my teeth on edge, as they say, and it still does when I see them doing it on television. What strange, inconsequential things occupy vital space on those memory cells of ours!

My recollections of the lectures and lecturers are largely anecdotal. There was the professor of Applied Mathematics, Dr

Joliffe, with his shock of pure white hair, who, for most of the time (like Sir J J Thomson) used to lecture with his eyes closed. He was mentally wide-awake though. I have never forgotten a remark he wrote against one of my examination answers. The question required one to find the point of impact on an inclined plane of an artillery shell, fired from the foot of the plane. I had attempted to use coordinate geometry, using the equation for a parabola for the path of the projectile, and that of a straight line for the inclined plane, the point of impact being found as the coordinates of the point which satisfied both equations. 'Simple, my dear Watson,' (I thought). But I got into an awful muddle with it, and failed to arrive at an answer. Professor Joliffe's comment was very much to the point., 'You would do better if you were content with humbler solutions.'

    The other remark on an examination paper that I have never forgotten was made by Professor Appleton, and showed equal astuteness. There was a question to which I didn't think I would be able to give a full answer, so I had left it till last, written what I could, and I believe I even stopped in mid-sentence to create the impression (I hoped) that I had been 'beaten by the bell'. What deviousness! But you can't pull the wool over the eyes of an eminent Professor of Physics. He wrote, simply, 'Was is time or knowledge that ran out?'

    Then there was Dr. Flint, professor of Theoretical Physics. He was very elegant in speech, dress, and manner, and his treatment of any particular topic, however intricate, was always very rounded, and often lightened by an amusing anecdote or two. My particular favourite came at the end of a demonstration of the frequency range covered by our hearing. He had used an oscillator, starting from the lowest frequencies, and moving steadily upwards. We had been asked to start with our hands up, and to lower them when we ceased to be able to hear the oscillator. Dr Flint told us of a student he had once had in his class, who was one of those people who could always see something, hear something, feel something, or whatever, long after the rest of the class had ceased to be able to do so. Said Dr Flint, 'On the occasion of the hearing test, with Clever Dick's hand still in the air long after all the other hands had been lowered, it gave me great pleasure to inform him that the hearing range of the higher primates extended far beyond that of Man' - adding that he had had no further trouble with the student...

On the whole, all the lecturers were good-humoured and competent – all except one, that is, whose name escapes me, though I can still see him clearly in my mind's eye. Dark, wavy-haired, young-middle-aged, bespectacled and studious-looking, he used to enter the lecture room, put down his voluminous lecture notes, find the place, and then proceed to read from his notes, verbatim!

I find it little short of astonishing, how many of the lecturers I can envisage so plainly still, after more than seventy years – even their names still coming to mind. Perhaps it was indeed because they all had such strong personalities – that is, all but that last one.

The bulk of the research of the Physics Department was concerned with atmospheric electricity and the propagation of radio waves, and it is almost literally true that everything stopped for a thunderstorm. On the roof of the College, there was a whole range of devices to do with the measurement of electric field strength and so forth, and if the lecturer was involved with this research, he would tender a swift apology, and take himself off to the roof to attend to his precious instruments - definitely not the place to be in a thunderstorm, but what devotion to the cause!

Other equipment on the roof was concerned with the measurement of the height and electric charge density of the various electrically-charged layers in the upper atmosphere. This latter research was directed by Professor Appleton himself, and his pioneer work in this field was recognized by having one such layer named after him – 'The Appleton Layer'. His research may sound somewhat obscure, but it came to affect the lives of everyone, since it provided the theoretical basis for the optimization of radio wave propagation.

Looking back some seventy years, I have realized that when I entered the College in 1932, public transmission of radio waves (commonly known as 'wireless' waves in those days) was barely ten years old, yet my recollection is that by then the majority of households had a 'wireless', although, with their 'bright emitter' valves and heavy lead 'accumulators' to feed their hungry filaments, they were incredibly crude compared with the modern transistorized and 'digital' radios. Well do I remember as a youngster taking the accumulator to a local garage every week or so, to get it 're-charged'. I think I got tuppence (two pennies) for my pains – a handsome reward.

These thoughts on the development of the household 'radio' have set me thinking about how many other major scientific discoveries or inventions have taken place in my lifetime. Aeroplanes and motorcars were in evidence before I was, but their mass production and popularization have certainly come about since I arrived on the scene. It can hardly be said that television followed hard on the heels of the 'wireless', since it was not until 1926 that the first public demonstration of a very crude system took place, the picture (black and white, of course) consisting of a mere 30 'lines'. It was another 12 years before a public television service (using 240, or alternatively, 405 lines) was inaugurated – to be closed down a year later by the outbreak of the Second World War.

What, then, of all the other innovations during my lifetime? The list is a startling one, and serves to underline the contention that a veritable explosion of discovery and inventiveness took place in the twentieth century. Thus, as we have seen, there was the mass production and popularization of the motorcar and aeroplane, which, long before the end of the century, had revolutionized transport, just as radio and television inexorably revolutionized communications during the course of the century. Less spectacular, but with far-reaching consequences nevertheless, were the fundamental advances made in the science of lighting and illumination. I can just remember streets lit by gas lamps that had to be turned on manually, and individually, by a man (a 'lamplighter') equipped with a pole with a hook on the end of it which engaged with a ring on the end of a chain attached to the gas tap. A second chain and ring enabled him to turn it off again in the morning. Compare that with the high-powered fluorescent lamps of today, switched on and off by a photocell sensing the ambient light level...

Not everybody rejoices in this particular development: my son John, an extremely enthusiastic and highly-skilled amateur astronomer, can often be heard leading off alarmingly about the evils of what has come to be known as 'light pollution'. And one doesn't have to be an astronomer to suffer from the consequences of *that*: the lives of all of us city dwellers has been impoverished by it, deprived as we are of the pristine glory of the night sky, awash now with the spillage from street lighting which is doing its best to turn night into day.

I can remember the gaslights in our house, too, with their 'incandescent mantles' which shed a cold, comfortless light, and regularly 'plopped'. Well do I remember, too, when my parents decided to be among the first in our little town to 'go electric'. A crop of poles appeared everywhere along the streets, draped between which were heavy cables, supported by large porcelain insulators. The cables could be traced back to the power station, which I can still see so clearly in my mind's eye, with the loud-humming dynamos in one huge shed, alongside which was another, filled with huge lead accumulators, each of approximately 2 volts, with sufficient of them strung together to provide something over 200 volts as the supply. Yes! It was *'DC'* – and particularly dangerous from the point of view of electric shocks, as muscles could go into an unremitting spasm.

However, for the novelty-struck first-time users it was a case of ignorance being monumentally blissful. We were down the garden on a summer's evening. I must have been around 9 or 10 at the time, and as we turned towards the house in the gathering dusk I remember running ahead, shouting gleefully as I went, 'Bags being the first to turn on the light!' But it didn't stop at that. Even at that tender age the urge to experiment was showing itself. There was the time when I pushed two fingers into a wall socket – fortunately the fingers belonged to the same hand, or the experiment might have led to my early demise. This experiment was shortly followed by another – that of trying to light up a torch bulb (I thought it would simply be extra bright) by putting it across the mains. The resulting explosive arc blew a fragment off the surrounding porcelain insulator, and evoked an almost equally explosive response from my father: I must have led a charmed life in those days.

The benefits stemming from the introduction of mains electricity into the home didn't stop at improved lighting. Over my lifetime we have seen mains-operated radios, 'gramophones' and television sets become virtually universal, together with a vast range of domestic equipment such as electric cookers - both conventional and microwave - refrigerators, freezers, washing machines, dishwashers and vacuum cleaners, to say nothing of a multitude of smaller devices such as food processors. Bear in mind in this context that I can remember my mother boiling up clothes in a built-in 'copper' in the scullery, having a fire under it, and subsequently putting the clothes through a 'mangle' to squeeze the water out of them. Indeed, I used to volunteer to turn the mangle handle – and hard work it was, too. Whilst on the subject of handle-turning, another one I used to turn regularly for my mother was the one belonging to her Frister and Rossman sewing machine. It was an object of some beauty, with its black-japanned finish, decorated with elegant gold patterns (would I have remembered it so well otherwise?). And how fascinated I was by its complex internal anatomy! – intricate cranks and oscillating bobbin arms revealed by sliding open strategically placed covers - like those cutaway diagrams of human anatomy exposing, all unsuspected, a complete heart, or a pair of lungs beneath the bland surface of the body's exterior. Of course, mangles have long since been superceded by 'spin dryers', which, together with sewing machines and a host of other household equipment, are all electrically driven.

After radio and television, we must give recorded music

rather more than an honourable mention, considering the role it came to occupy in our daily lives. At the time I was born it was, I believe, recorded mechanically on a rotating 'master' disc by a steel needle attached to a vibrating diaphragm that was itself mounted on the end of a large horn, whose job was to collect enough energy from the sound waves in the air to produce an adequate amplitude of vibration of the recording needle. Copies of the master disc could then be 'imprinted' on blank discs made of suitable material. (The playing time of these discs was approximately 5 minutes per side!) This crude system was superceded in the late 1920's by a system using a microphone coupled to a magnetic 'recording head' via an electronic amplifier, a system which, with improvements and refinements (such as the 'long play' records introduced after the Second World War, and 'stereo' records in the 1960's) lasted until the beginning of the 1980's, when a giant leap forward was made by the introduction of the 'compact disc', or CD. This provided over twice the recording time on a disc less than half the diameter of a 'long-play' record, and with near-perfect fidelity. Looking back on the development of sound recording, which occupied the greater part of my lifetime, I am amazed that any system based on mechanically engraving a disc worked as well as it did, and that it began to be superceded only a couple of decades or so ago.

It is fascinating to realize that those same two decades (1980 to 2000+) saw the arrival of the personal computer, and the whole of its development into the extraordinarily sophisticated instrument that it now is. The mind boggles at what it will have developed into, in another two decades. My grandson Kevin, I'm sure, can hardly wait to find out...

Of course, behind the phenomenally rapid development of the computer, sound recording and radio and television, lay the development of the transistor, and the 'microchip'. It is fascinating to recall that the basic concept of the transistor, like many another epoch-making development in science, had its beginnings in a chance observation made during measurements totally unrelated to such a concept. As in other chance observations made during the course of scientific experimentation, the vital step was to ask the right question at the time: 'Why did that happen?' and (equally important), 'Does it have a use or application beyond the scope of the present investigations?' These vital questions were followed up after it was observed - during measurements in the early 1950's on a so-called 'semiconductor' - that a small change in the voltage of one of the electrodes attached to the semiconductor produced a relatively large change in the current carried by another electrode. And so the notion of the transistor was born. Another famous example was the discovery of penicillin as the first in a whole new class of substances that came to be known as 'antibiotics', following the observation that a mould growing on a culture medium used in the study of bacteria inhibited their growth. Many more examples could be quoted, but perhaps a humble one from my own experience will bring it down to the personal level. I detached a wire from the terminal on a piece of equipment, and a meter jumped to a new reading, and simply asking myself why it had done so, led to an extremely accurate method of calibrating the equipment!

But I digress, and to return to my catalogue of epoch-making developments during the twentieth century, I cannot fail to mention Man's ventures into space, most notable of all, of course, the moon landings. The exploratory missions to other planets, such as Mars, by means of so-called space probes, are hardly less spectacular, with the enormous distances involved, and their ingenious 'buggies' trundling about the planet's surface, looking for evidence of life, past or present. Why do I refer to the exploration of space as requiring 'at least' a 'mention' then? I haven't seen the sums, but there is no doubt that the amounts of money spent on it have been vast. So why just a mention? The fact is that it has had precious little impact on life on planet Earth, despite the vast expenditure, and it has been dogged all along by repeated questions as to whether the money could have been much better spent on Earth rather than in Space. There has been relatively little technological spin-off, either, to offset such expenditure. The only item that comes to mind is an 'orthopaedic' mattress, said to be made of the same material as an astronaut's couch. I have to say, however, that despite the criticism, one has found oneself glued to the television at critical times, notably during the first moon landing. I suppose it is a fact that such is Man's nature that with astronomical space all around us, it was only a matter of time before he would venture into it, whatever the cost. It's the mountaineer's response to 'Why climb Everest?', isn't it? Incidentally, I almost overlooked that, didn't I? It was 1953 – just days before Queen Elizabeth the Second's Coronation. How it stands out amidst all the technological achievements with which I have inevitably been preoccupied! The pitting of a few men's bodily strength and powers of endurance against everything that the highest mountain peak on Earth could throw at them stands out refreshingly against the voluminous catalogue of scientific advances. Brawn or brain, one might say, but the elemental nature of the conquest of Everest is in many ways more inspiring than the conquest of the Moon, which required vast teams of men, and the expenditure of vast sums of money for its success.

For the penultimate item on the list of inventions and discoveries in the twentieth century which have turned out to be of prime significance, I have chosen the elucidation of the structure of deoxyribonucleic acid (DNA) in the early 1950's, with all that has stemmed from it in the field of genetics, and will continue to stem from it for the foreseeable future. This is not my field, but I think even 'the man in the street' is at least vaguely aware of the challenges implicit in so-called 'genetic modification', together with (for example) the unravelling of the hereditary aspects of many diseases, and the forensic use of DNA matching.

At the end of my list, but top of it in terms of its influence on the course of the history of the twentieth century, and its terminal consequences for the whole of mankind if it is ever used in anger again, I must place the atomic bomb, in all its shapes and forms.

It began, just before the turn of the century, with the

discovery of the electron by Sir J J Thomson , followed by the discovery of X-rays by Röntgen, and radioactivity by Becquerel, hard on the heels of whose work came the work of Pierre and Marie Curie on the radioactive ore pitchblende, from which they isolated the intensely radioactive element radium. The discovery of X-rays and radioactivity ushered in a whole new era in medicine (and, on a personal note, it was reading the biography of Madame Curie by her daughter Eve in 1943 that changed the course of my life – see later).

Following up Becquerel's work, Rutherford demonstrated that the radiation emitted during radioactivity had three components which he called, respectively, alpha, beta, and gamma rays; and later, at Cambridge, in 1919, he bombarded nitrogen gas with alpha particles and produced an isotope of oxygen and sub-atomic particles called protons – the first artificially induced nuclear reaction, a discovery which led to awesome consequences when, during the Second World War, the first atomic bomb was devised. This followed the realization of the possibility of a sub-atomic so-called 'chain reaction' – which, in fact, brings us back to my university days.

In my very first term at King's, there came a morning when one of the lecturers (his name escapes me) walked into the lecture room with a huge grin on his face, put down his folder of notes, and announced that 'We've found the neutron.' (The 'we' was used collectively, of the profession of physicist, and not personally.) Its existence had been predicted as early as 1919 by Rutherford and others, but it was left to Chadwick, working in Cambridge in 1932, to deduce that the results of experiments by French physicists confirmed its existence.

It is an uncharged particle of mass virtually equal to that of the positively-charged proton, and, like the proton, is a fundamental component of all atomic nuclei - apart from ordinary hydrogen.

Yes, our lecturer was *very* excited by the discovery, and rightly so, because it is one of the fundamental sub-atomic components of matter. But I wonder if his response to the news would have been adequately described as merely 'excited', had he been aware of the possibility of the so-called neutron chain reaction - the basis of the atomic bomb? How could he know that the discovery - of mere academic interest at the time - would lead to the design of a fiendish device that, a mere 13 years later, would kill or injure 130,000 people, literally in the twinkling of an eye? I am sure that, had he known, it would certainly have wiped the grin off his face.

Oops! – I nearly forgot plastics! They did as much, and perhaps more than anything else I have mentioned, to change the life of the ordinary man. Bakelite, developed in 1906 was followed, through the century, by an almost endless list: rayon, nylon, PVC, perspex, polystyrene, PTFE, teflon (immortalized by its use in non-stick cooking pans), polyethylene, etc., etc. And what about the role of electronics in the development of a whole range of musical instruments, from the electric guitar to electronic pianos and church organs, virtually indistinguishable from the real thing? That particular oversight of mine is almost unforgivable, considering the role of music in my life.

But I must return now to the event that was the culmination of my undergraduate years – the Final Examination in Physics. It took place in June 1935, a month that also marked my 21st birthday. Immediately, I find myself comparing my attainments during those first 21 years of my life with those of the last 21! But then, they do say that comparisons are odious...

That Final Examination consisted of 6 three-hour written papers, together with a Practical Examination. On three consecutive mornings, I had to get up at around 6.30 AM in order to catch a train that would make sure of my arriving in good time for the 10 o'clock start to the first of the two papers set for that day. For almost all my working days I commuted to and from work, and for one reason or another (including the time when there were 'buzz bombs' overhead) many of them were fraught, but I think those three Finals' days have to rank as the worst. And despite the early start, well do I remember on the night before the last of them going for a walk at 1 AM with my future brother-in-law. It was a cloudless, moonlit night, its clarity matching the clarity of the arguments with which I sought to convince him, with all the persuasivenes of a condemned man, that I was going to fail. I know now that this is a common phenomenon, but that night my brother-in-law's task was nothing less than to convince me that it was worth my while catching that train the next morning, in order to take the last two papers. His argument that I had nothing to lose and everything to gain finally won me over, and five hours or so later, I was getting up again to keep my appointment with the examiners.

How does one get it so wrong at times? I knew that I had worked very hard indeed those last two years - if for no other reason than that of repaying my parents for all the sacrifices they had made to enable me to go to University at all. The reason may have been epitomized in the title of a radio play I heard round about that time. It was called 'The Flowers Are Not for You to Pick'. I don't remember anything about the play itself, but its title has stuck in my mind ever since. Why? I think that all that chronic illness and hypochondria with which I was surrounded all my childhood had subconsciously conditioned me into believing that it was too much to expect happy outcomes – something nasty always intervened to deprive you of it - better not to expect anything really nice to happen, then you won't be disappointed: the flowers are not for you to pick.

Well – I was wrong, *massively* wrong on this occasion. All that hard work had paid off: I got a 'First'! My memory of how I came to learn that is very vivid, too. It was not through the post - that must have come later. The results were posted up outside Imperial College, I think it was. I just could not contemplate going on my own to find the dreaded truth, so my father came with me, and the scene outside the College is etched on my mind. There were already groups of students standing around talking excitedly, but there was a lone student walking away from the board with a haunted look on his face, which I have never forgotten. He was a fellow Physics student from King's. 'Somersby!' I exclaimed - but he walked on in silence. And when

I got to the board myself, I realized why – his name did not feature on it at all. What salt my 'First' must have rubbed into that gaping wound of his...

I have just recollected an amusing incident stemming from the exam result. It was still early evening when my father and I arrived back in Stevenage – and, all agog, I felt I just had to go and tell my old Headmaster, Mr Thorne. Despite the fact that I was the bearer of good news, it was with some trepidation that I knocked on the front door of his house – the first time I had ever done so. My misgivings seemed amply justified: he must have been standing the other side of the door with his wife - hats on - and at that very moment about to go out.
'What *is* it, Kemp?' he asked tetchily. 'We are just off to the *cinema*.'
'Well, you see, sir, it's my exam result – I got a First.'
'Come *in!* Come *in! Splendid!* Well *done!*' he effused, as he ushered me into his study, his wife completely sidelined, and the cinema forgotten.

I had applied for, and been granted, a Bursary for a fourth year, to study for the Postgraduate Diploma in Education at The Institute of Education, my intention, at that time, being to take up teaching. Despite the fact that as things turned out I made only a very limited use of it subsequently (altogether about four years in full-time teaching), I have never regretted that fourth year, or regarded it as marking time, or, worse still, as a waste of time. From the point of view of my own education, in the broadest sense, a Special Honours Physics degree, with subsidiary Mathematics, could hardly lay claim to being rounded, any more than could, say, a degree in Philosophy, with no understanding of, or insight into the ways of modern science. Among the subjects I studied in that fourth year were the history of science, teaching methods, and educational psychology, as well as a course which included human anatomy and physiology, and must have come under the general heading of Health Education. My mental horizons were certainly much broader than they would have been had I left the University at the end of my Degree course.

There was trouble ahead, however. Once I had started the Education course, I paid a visit to the Research Laboratories at King's to see what a handful of my former classmates were getting up to, in their M. Sc and Ph. D courses. It was nearly my undoing. I found myself green with envy. It was Merriman that I remember so clearly: he was investigating the behaviour of a tetrode valve. All those knobs and dials, and the esoteric graphs he was plotting! – 'Surely this is where I belong', I said to myself: breaking new ground, and not in the classroom teaching virtually the same old material, year after year. 'Somebody's got to do it', I reasoned with myself. But I knew enough about teaching by then to know that it had to be done with conviction and enthusiasm if it was not to be a waste of time for you, and your pupils. However, I was committed to that fourth year, and, as I have already said, glad of its broadening influence, but I knew in my heart of hearts that I could not now make a career of teaching.

At the end of the year, I duly obtained my Diploma. One of the things I can't remember is whether there was an understanding that I would teach for at least two years. I suspect there was. Anyway, I was 'up front' with the Department of Education about my feelings, and they showed a welcome understanding of my position.

I have but a handful of memories of the Diploma course itself. The first one that comes to mind is a somewhat dramatic one, of an incident in one of those Health Education lectures. Remember – we weren't case-hardened medical students, and an elderly medic – a lady – had brought along a 'pickled' human brain to be passed round the class, to enable us individually to inspect some lesion or other which had caused the death of its owner. I scrutinized it rather squeamishly, and passed it on to my neighbour. He took one look at it and passed out on the floor. (The lecturer did not waste the opportunity to demonstrate her first aid skills!)

One could say that in general the nature of the material taught under such headings as 'Teaching Methods' and 'Educational Psychology' was much more likely to provoke debate - and evoke differences of opinion - than the material comprising a Physics lecture, especially when fundamental differences in approach between lecturers emerged from time to time. It was, in fact, inevitable, with one of the lecturers (Mr Beales), who taught History of Science, and was a devout Catholic, and another (one Dr Titley) whose subject was Teaching Methods, who was an evangelizing atheist. I have a vivid memory of an incident following one of Dr Titley's lectures. I cannot now think how he had managed to provoke me into a response on the subject of religion (I use the word a little reluctantly – I would have preferred 'spirituality'). He did, however, and I left a terse little note on his desk to the effect that his position was tenable only if he was content to live his life in terms of what could be measured and/or weighed. Imagine my surprise and, indeed, embarrassment when *Mr Beales* quietly said to me at the end of one of his lectures, 'I hear that you have been crossing swords with Dr Titley.'

It was, indeed, a stimulating year, and provided opportunities – particularly in Mr Beales' lectures – for airing one's concerns about matters running a great deal deeper even than sub-atomic physics.

I must not give the impression that there were no such opportunities during those first three years. King's had a strong Theological Department rooted in its foundation articles, in which the purpose of the College was stated to be that of 'giving instruction in the various branches of literature and science and the doctrines and duties as the same are inculcated by the United Church of England and Ireland.' Although a somewhat convoluted statement, in short it amounted to adding the teaching of 'the doctrines and duties inculcated by the United Church of England and Ireland' to the normal duties of a university college. Those who attended the Theological Department with a view to taking Holy Orders became a *theological* 'Associate of King's College', or, in short a theological AKC. However, the opportunity of 'instruction in... the doctrines and duties of the Church...', was extended to *all* students by the provision of a so-called *non*-theological AKC, which was awarded by examination, after attending a two-term course in Theology (one lecture a week). 'Non-theological AKC' was something of a contradiction in terms – but be that as it may - that two-term course I well remember as highly stimulating, providing a welcome change from the matter-of-factness of physics and mathematics.

There was one further activity comprising an essential part of the Diploma course – namely Teaching Practice. I did mine at a well-known school in south-east London, the name of which escapes me now. The senior Physics master was a kindly and helpful soul, and had a good way with words when he wished to make a point. So much so that the only clear memory of my time at the school that comes back to me is a comment he made after sitting in on a lesson of mine. 'The trouble is,' he said, 'that inserting my presence into your teaching period is rather like putting an ammeter into a circuit to measure the current in it – the ammeter disturbs what you're trying to measure.' How apt - and how succinctly it makes the point! And that was Mr Nicholls, another name to come readily to mind after nearly 70 years.

As I have said, I valued that fourth year for its contribution to my general education, but, in point of fact, it was also to stand me in good stead only a few years later.

Chapter 5

It might be thought that when I finally emerged from University in June 1936 with a First in Physics and a Postgraduate Diploma in Education (to say nothing of a 'non-theological' Associateship of King's College...) I would have been justified in feeling that the world was my oyster. I felt no such thing. Not only the national economy, but also the world economy was far from healthy, and, what is more, throughout my days at the University the storm clouds had been gathering, as Hitler's Germany began to threaten world peace.

From a young age I had attended Sunday School at the Baptist Church, and had become a regular attender at the Church itself. To complicate my involvement with the Church writ large, I had started to help out with the Sunday School at the Methodist Church in Knebworth (a ten-minute cycle ride away), where my old school friend Clifford Walker was involved. This eventually led to my leaving the Baptist Church, and transferring my allegiance to Stevenage Methodist Church, which was in charge of an exciting young minister named - believe it or not - Joseph Heaven, and very soon I had become the Superintendent of the Sunday School. It is plain that I took my Christianity seriously, and finding myself unable to believe that there was any such thing as a 'just' war, I had joined the Peace Pledge Union, a country-wide pacifist organization which still exists.

No, it's true – there were no 'pearls' to be snapped up in the world I was entering. It was ominous and threatening, and even though I was in possession of that first-class degree, there weren't all that number of suitable jobs to be had, with many of them in scientific research already beginning to have connections with the arms industry, and, as far as I was concerned, not therefore an option, thus narrowing the field considerably. Teaching would have been an obvious choice, but I had already ruled that out. So what? Ah -there was television. Though no public service existed, earlier versions of the present-day system, based on the cathode ray tube, were in an advanced state of development by companies such as the General Electric Company and Marconi. At the time, I could see no possible connection with war-making, and within weeks of leaving College I had applied for, and obtained, a post at the Research Laboratories of The General Electric Company, in North Wembley.

I was interviewed by one Dr C C Paterson, a well-known personality in the world of industrial science, and I have to say that he was somewhat unhappy about that year I had spent at The Institute of Education. He made it plain that he regarded it as a 'wasted' year, and I think he felt that it would have taken some of the polish off my first-class degree. He was probably right about that, and in any case, as a physicist, I soon found myself feeling something of a fish out of water among my workmates, nearly all of whom had degrees in electrical engineering, and consequently had a much more intimate working knowledge of electronic circuit design than I had, as a 'pure' physicist. Even so, we were under extreme pressure to produce a working television, capable of mass production, by the time of the 'Radio Show', held annually, in the Autumn. I found myself spending most of my time with a coil-winding machine, producing radio-frequency transformers, used for coupling one valve stage to another. The work was totally empirical, repetitive, and very boring, and could have been done by any intelligent sixth-former. I had, in fact, made a serious mistake in my choice of that first job: I was as mis-matched to it as two of those valve stages would have been, with a badly designed inter-stage transformer. The physics - as such - involved, was minimal and very elementary, and I rapidly lost touch with most of what I had been taught at King's. In fact, it could easily have been disastrous for me.

For the time being, though, I had to earn my keep, and put my back into the job with the rest of my workmates. I was, incidentally, still living at home in Stevenage, and commuting daily to North Wembley. This was bordering on the ridiculous, getting up at 6 o'clock in the morning to clock in at 9 am in North Wembley, not arriving home again until around 8 o'clock in the evening. It went beyond a joke a few nights before the Radio Show, when it was deemed necessary that we should work all night. This meant getting up at 6 o'clock one morning, travelling to North Wembley, working all that day and night, and all the next day. At 4 o'clock that afternoon my nervous energy ran out, everything about me seemed to recede to a distance, and I asked leave to 'call it a day', or rather 'two days and a night'. When I got home I went straight to bed, and woke only once in a twelve-hour sleep. Why should I remember that? – simply because I started to drink some water out of the glass by my bed, and went to sleep again before I had put the glass down. Need I say more?!

There were bye-products of the research work in the Television Laboratory - namely the prototype television receivers, which, in turn, were made redundant by on-going improvements in design. And these, instead of ending up on the scrap-heap, became available to staff, on a semi-permanent basis. No big deal, you might say. Well. just think – the receiver I acquired in 1938, and set up in my parents' home in Stevenage was, as far as I know, the only one in a town of 5000 inhabitants. The upshot was that almost overnight 43, Basils Road became more like Piccadilly Circus, with people continually coming and going to get their first glimpse of this latest miracle of science. Strangely, my parents seemed to take it all in their stride, and, I think, rather enjoying the limelight!

The 'Sunday Play' was the climax of their week, when they were hosts to something like a dozen neighbours, friends, and special guests. Among the latter was the catholic priest (just imagine – with my mother brought up on primitive Methodism!). The only plausible explanation I can think of is that the manse was also in Basils Road, which my parents might have thought entitled him to automatic membership of the 'Television Viewers' Club'... Incidentally, my mother, responding to the informal friendliness of these occasions, provided sandwiches all round, week upon week, thereby adding to the communal spirit of the occasions.

Even after seventy years, I can still recall the truly child-like (no! – not child-*ish)* excitement of those Sunday evenings, and inevitably it has set me pondering the fact that such experiences are virtually a thing of the past. Nothing surprises us any more, and, having lost our sense of wonder, we have become blasé, and our lives consequently impoverished. How grateful I am to have been born at a time when to talk to someone on the other side of the world *within a few seconds of deciding to do so* was unimaginable! And as for *seeing* something happening in Australia *as it is actually happening!* - such a suggestion would have been deemed the purest-of-the-pure science fiction. Present-day children, born into a world in which all these things are fact not fiction, take the continuing near-miracles wrought by modern technology for granted, so that even the developments that have taken place in their *own* life-times - such as computers, the Internet, and mobile phones - hardly cause them to bat an eyelid. *What* a pity!

There is one more striking example I must record. We are all familiar with the fact that the picture we see on our televisions is created on the screen of a so-called cathode-ray tube, which most people will assume was created for that purpose, but that is not so. Long before the cathode-ray tube became an integral part of a television receiver it was being employed as the basic component in a so-called 'oscilloscope', a laboratory instrument used to display - in graphical form - the fluctuations in, for example, an electrical voltage (the fluctuations, if regular, being called the 'wave form' of the electrical quantity). Remember then, that before I took up my post at the Television Laboratory I myself had never seen anything else on the end of a cathode-ray tube *other than* such graphs or waveforms.

Now, since there was no public signal on which we could test our receivers, we had to generate our own signal for this purpose, which was not broadcast, but fed to the receiver under test by means of a so-called coaxial cable (with which, of course, all of us are familiar, these days!). The test signal itself was generated by a simple loop of film, maybe of the order of ten feet in length – in fact a miniature 'movie' lasting just a few seconds, and repeated, endlessly. So what was so special about that? Well, at the GEC laboratories it was of a huge 'shire' horse, out in the middle of a field on a windy day, with its long mane flowing freely, in the wind. How can I convey the sheer *magic* of seeing – *in place of a mere graph* – the gorgeously long mane of a horse, tossed hither and thither in a wind 'blowing where it listeth'? You don't get what I'm on about? No – I feared you might not. But doesn't that prove my case? – familiarity may not always breed *contempt,* but, alas, it does so often these days breed *indifference,* which can be almost as bad.

Be that as it may: some photos now. Plate 4 is of myself, inspecting that loop of film just mentioned, for possible damage arising from its endless transits through the projector, whilst Plates 5a and 5b are, indeed, *historic.* They are photographs I took of the picture displayed on the cathode-ray screen of an actual receiver, during some of the earliest public transmissions, in 1938. Plate 5a is of Elizabeth Cowell, one of the two earliest women TV announcers: she became a household name. Note the curved sides of the picture: the screen was little more than six inches in diameter, and one had to make the most of it – at the cost of dispensing with a strictly rectangular picture (beggars can't be choosers!). Plate 5b is of a scene in a TV drama (televised 'live', in those days).

To return now to my main story-line: after the accident with the glass of water, I decided that there was nothing for it but to get digs near the Labs. I was lucky. I quickly found a comfortable bed-sitter, with breakfast and evening meal, and a few minutes' only, from work. What a luxury it seemed, after all that commuting! My landlord and landlady, a Mr and Mrs Watterson, were a late middle-aged couple from the north, evidencing the north's traditional, no-nonsense hospitality, *and* Mrs Watterson's excellent cooking!

Mr Watterson was an accomplised pianist, and I can still remember his bravura rendering of Sinding's 'Rustle of Spring'. I realized how fortunate I had been in my choice of digs – unexpectedly and surprisingly so. From a very young age music

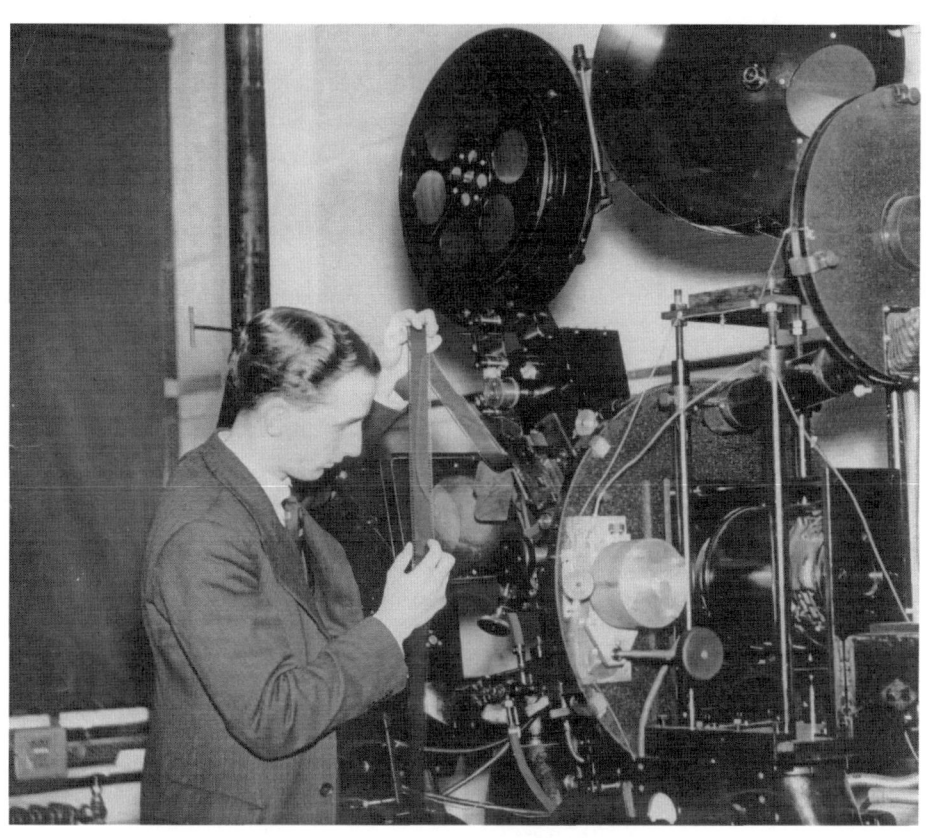

Plate 4 Inspecting the TV film loop

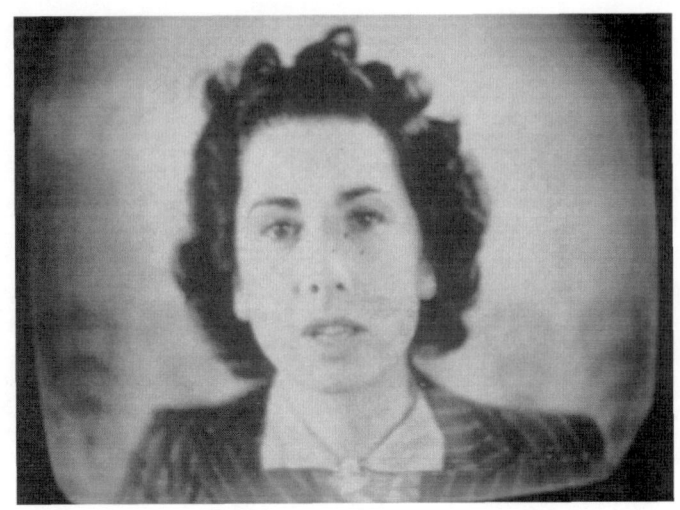

Plate 5a Elizabeth Cowell, TV Announcer, 1938

Plate 5b Live TV drama, 1938

had been a great love of mine. Long before I could read music I was drawing lines on a piece of paper and putting black blobs on them – my first attempt at satisfying what came to be a life-long, burning desire to be a composer! I had a few piano lessons around the age of 10, but Grammar School and University combined to prevent music from occupying the place I would have liked it to have had in my life. But now, I reasoned, as I listened to 'The Rustle of Spring', if it was OK for me to listen to him playing the piano, it must be in order for him (and his wife) to listen to me, doing likewise. So – having, of course, asked permission - I duly bought a piano, and installed it in my digs!

It was a 'Monington and Weston' double-iron-frame upright. Before I bought it, I visited the factory in north-east London where they were made, and they proudly told me that its construction provided such stability that they could ship pianos to China, and they were still in tune on arrival! I bought mine new in 1937 for £36! Impossible? No. That was about a sixth or seventh of my annual salary as it then was, in my first job. And that's about what a similar piano would cost today – say £2500 or so. That selfsame piano is just across the room from me as I write – and it's almost as good as ever.

I started having piano lessons again, but a chance discovery soon relegated them to a firm second place in my musical life. I had found that in the very next road there lived one Dr Eric Thiman, a well-known organist and composer who gave composition lessons. I approached him somewhat gingerly. After all, my physics degree and Postgraduate Diploma in Education (to say nothing of my non-theological AKC...) were hardly relevant to the composition of music, or any evidence of my ability to do so.

Dr Thiman took me on though, and I suppose that, in all, I had about a couple of dozen lessons from him. It was an idyllic time for me. Just think! There was I, a physicist, working as a total misfit in a laboratory whose business was electrical engineering, having lessons in composition. It was the time of my life! My physics, for the time being at any rate, receded still further, as I learned to think music in my head, commit it to paper, and then convert it to sound on that new piano of mine. At first it was little gavottes and walzes and the like, but the church musician soon had me setting words to music, and I felt as proud as if I'd discovered another sub-atomic particle when he told me I had a gift for it − me! − a winder of coils on cardboard cylinders had a gift for setting words to music... Yes, it was indeed an idyllic time for me. But it didn't last. Somehow I thought it wouldn't − *couldn't*, really. I was too happy − remember? − 'The flowers are not for you to pick'...

It was 1938, and what came to be called simply 'Munich' filled the headlines. My Peace Pledge Union activities had intensified accordingly, and I told Dr Thiman one day that learning to write music in the kind of world in which we were living was a bit like fiddling while Rome burned. It was neither wise to say such a thing (nor, actually, was it true). He was livid. And the next − and only − time I saw him again (it was in a restaurant in Wembley) he 'cut me dead'. How it hurt! − the man I had felt so close to, as we worked together on what I loved even more than my science. How could he do that? On the other hand, how could I have implied that music was irrelevant in a warring world? What a traitor to the cause he must have thought I was.

Despite the gathering war clouds, 1938 was a very happy one for me, simply because it was in that year that I met Mary, who, three years later, was to become my wife: Plate 6 is a photograph I took of her at the time of our engagement. The manner of our meeting was both serendipitous and convoluted. My parents went on holiday to Ventnor, in the Isle of Wight, whilst Clifford Walker and I went on a walking holiday in the Lake District. It so happened that the hotel porter took my parents' suitcases to the wrong bedroom − one occupied by two young ladies from Wanstead, in Essex - the suitcases belonging to the two young ladies, by the law of reciprocity ending up in my parents' room. In the process of sorting out the muddle, the four of them became quite friendly, and spent a good deal of

Plate 6 Mary, in 1938

their holiday time together - so much so, that in the Autumn of 1938 my parents invited the two – Kathleen and Mary – down to Stevenage for the weekend. Soon afterwards, Mary and I met up in London and went to see 'Snow White and the Seven Dwarfs,' at that time receiving its premier showing in this country. And soon after that Mary was coming to Stevenage on her own for the weekend, and we were enjoying long walks together in the beautiful Hertfordshire countryside.

The times were hardly idyllic, and the courtship in consequence far from carefree - Mary was as keen as I was in support of the anti-war activities of the Peace Pledge Union. There was also The Fellowship of Reconciliation, a specifically Christian organization, in which we were both active. Membership of such organizations amounted to only a tiny fraction of the total population, and one faced a great deal of hostility in consequence. Two particularly unpleasant incidents I well remember – one was being spat at, and the other was being passed in the street by one of my old Grammar School masters with whom I had been on very good terms while at school. He went by me as though he had never seen me before. He had been a Colonel in the 1914 – 1918 war, and I understood the strength of his feelings, but not the way in which he expressed them.

Through 1938 and into 1939 the sense of isolation increased steadily, and a small group of us began to study the situation from first principles. What was wrong with the Western way of life? What could create such stresses and strains within, and between, different societies as to cause them to resort to such a heinous method of settling their differences as modern warfare? What, by way of example, could we do? We had already done all the usual things – pamphleteering, writing letters to the press, holding public meetings, and so on, but we no longer felt it was enough. We had to find, and demonstrate a way of life that would remove the root causes of war. All that one small group of half a dozen or so could hope to do was to serve as a sort of demonstration model of what could be done if the will was there.

What emerged (in 1940) was 'The Stevenage Simpler Living Group', which, to many, to quote St. Paul's word from another context, might appear utter 'foolishness'. Basically, it required each individual member, within his or her own conscience, to examine their lifestyle, and to decide what he or she regarded as essential expenditure, and what was non-essential expenditure that could be put to better use by the group as a whole. The decision as to what was regarded as essential, and what was regarded as non-essential, was left to the individual. It was an understanding within the group that such decisions were not to be harshly ascetic ones: a modest allowance could be made for expenditure on, for example, books and gramophone records. It might be thought - quite understandably - that such 'loose' constraints would not produce a significant outcome. In point of fact, from its half a dozen or so members (one of whom was out of work) the group found itself with about £3 a week for funding its activities within the community. Not much? You're wrong. It would, I believe, have been the equivalent of around £300 a week these days. These group activities included going to local grocers and coal merchants to find out who their poorest customers were that would not say 'no' to some extra groceries each week, or an extra hundredweight of coal on occasion. Trivial? Not so, it would seem, from the 'feedback' we had from the grocers and coal merchants who participated in the scheme. The source of these extras was kept strictly anonymous – we were not seeking publicity, only a worthwhile outlet for our self-determined 'surplus'. We did, however, publicize our activities, by talking to Church groups and the like.

The Group had always regarded itself as loosely based on the so-called 'Third Order of St. Francis' – a lay Order dedicated among other things to working with their hands for others, and once war had broken out we soon found an opportunity to do just that. Stevenage was a designated area to which children from London were sent to escape possible bombing raids. Spare beds were soon at a premium, so the Group set to work to make hessian mattresses (I can't for the life of me remember what we stuffed them with – probably straw!). As work place we were loaned, free of charge, an empty shop, and over a period of months we turned out dozens of mattresses, gratefully received by households who had had evacuees dumped on them regardless of their accommodation – or lack of it.

So what happened to the Group? Alas – the war saw to it that in due course its members became scattered, if not to the four corners of the earth, then at least to the four corners of the country, and we quickly realized that the very nature of the Group was such that it required the personal presence of its members in the shared fellowship of its activities. As well, Mary and I had married, and we soon had our son John to consider. Would we have felt able to join another such group, had the opportunity arisen? I can't say. I really don't know. What I do know is that I look back on that time – just two years or so, it was – as one of the most spiritually fulfilling periods in my life, and the fact that it ceased so soon, and so abruptly, one of the greatest regrets of my life. Nevertheless, and despite those regrets, there is no doubt that those two years, despite their brevity, left their mark on me, and changed, for ever, my attitude to money, possessions, and acquisitiveness in general.

The principles underlying the formation and activities of the Group were embodied at the time in a small pamphlet of some two thousand words or so, entitled 'Christian Stewardship and Community', with the sub-title, 'An experiment in community living by the Stevenage Peace Group'. The text of it is reproduced, verbatim, in Appendix 1. Re-reading it after all these years has convinced me that it is just as relevant today as it ever was: just think of hundreds of small groups up and down the country, each with several hundred pounds a week to spend on good causes…

NATIONAL SERVICE (ARMED FORCES) ACTS

Local Tribunal for the Registration of Conscientious Objectors.

To Mr. L.A.W.E. Kemp,
43, Basils Road.
STEVENAGE. Herts.

59-62, Queens Gardens,
BAYSWATER. W. 2.

..................................(date)

Case No. E.1402A

NOTIFICATION OF RESULT OF APPLICATION

At the hearing of the above Tribunal on 30.10.40. your application for registration in the Register of Conscientious Objectors was considered and it was decided :—

* (a) That you shall, without conditions, be registered in the Register of Conscientious Objectors.

* (b) That you shall be conditionally registered in the Register of Conscientious Objectors until the end of the present emergency, the condition being that you must until that event undertake the work specified below (being work of a civil character and under civilian control) and, if directed by the Minister of Labour and National Service, undergo training provided or approved by the Minister to fit you for such work :—

that you continue in your present occupation, School Teacher; or engage in full time work on the land, or undertake whole time Civilian Ambulance Work.

* (c) That your name shall be removed from the Register of Conscientious Objectors and that you shall be registered as a person liable under the Acts to be called up for service but to be employed only in non-combatant duties.

* (d) That your name shall, without qualification, be removed from the Register of Conscientious Objectors.

A copy of the evidence before the Tribunal and their findings is enclosed.

An appeal may be made to the Appellate Tribunal within twenty-one days of the date of decision by forwarding a formal application on form N.S. 24, which may be obtained on application at the above address.

Signed..
Clerk to Tribunal.

*Strike out inappropriate items before issue.

N.S. 19
(5936-4937) Wt. 19190-4305 40,000 7/40 T.S. 677

Plate 7   Certificate of Registration as a Conscientious Objector

Chapter 6

War broke out on September 3rd, 1939, and I resigned from my job at the General Electric Company Laboratories immediately, since the Laboratories would be going straight over to work for the military machine - the particular laboratory I worked in probably to a top-secret project – radar. In October 1940 I appeared before a Tribunal to state the reasons why I was a 'conscientious objector', and was duly registered as such (see Plate 7). I immediately became something of a social outcast – particularly in my home town of Stevenage, and I had the difficult problem of finding work not associated with the war. The answer was, of course, teaching, and that ill-fated Postgraduate Diploma in Education was to prove useful after all. In fact, in a matter of weeks I was – to my surprise - working again. I obtained a post at The Regent's Street Polytechnic, just north of Oxford Circus. My main subject was, believe it or not, Building Science -   teaching things like the principle of siphonic action as applied to the flushing of toilets! Agreed, there was a little more to it than that, but it was yet another giant step away from pure physics, though, of course, it provided me with my daily bread. The Head of the Mathematics and Physics Department was one Dr Topping, who was a Quaker, and, as such, was fully understanding of my position regarding the war: he himself would have been a conscientious objector had he been of calling-up age.

In the event however, the job lasted only a few months. The Department was evacuated to Lancaster, a move which certainly didn't suit me personally: with Mary in Wanstead, and working in the City, we were able to meet up quite often, but it would have been another matter altogether if I were to move to Lancaster. The only work left behind by the Department involved teaching cadets bound for the armed forces, so, for a second time, I found myself resigning my job, and out of work. But once again, that fourth-year Diploma was to stand me in good stead.

One thing soon became obvious – that I was not going to

find another job in London, and that I would have to accept that - after all - the war was going to separate Mary and me, along with millions of other couples. It was late 1940, and I found myself applying for a job at a Grammar School in Bristol. I got it, and soon I was in digs again – in Congresbury, a village a few miles south of Bristol. It was at the home of the local chimney sweep, and a very nice man he was, too, with a nice, homely soul of a wife.

There had been no problem of discipline in the classroom at The Polytechnic, which counted as an adult institution, but it was to be a very different matter at that Grammar School in Bristol. It was a large co-educational establishment, and it took very few forays into the classroom for me to discover the difference. One incident comes to mind immediately. It was a very warm afternoon, and the sun was pouring in through a high window. 'Please, sir, may I pull the blind down?' 'Of course!' I said. And yes, the blind came down – roller as well, all of it, onto the floor. 'I didn't say you could do *that*,' I responded, angrily. 'But, sir, you did say I could pull the blind down'. I threatened to send the culprit to the headmaster, and an audible titter went round the class. It was obvious that I could expect no support from that direction. I soon discovered, too, that there had long since been what might be called an inflationary spiral in regard to the more mundane forms of punishment such as 'lines'. 25 to 100 would have been the norm, but such levels were scorned - even seeming to create disrespect for the imposer of such a modest reprimand: no - 500 was much more the going rate.

It was Autumn, 1940, and the so-called Phoney War was over, and Bristol was coming under air attack. During one of these raids there was an incident in Congresbury in the early hours, which had its amusing side. Already awake on account of anti-aircraft fire, I heard a heavy thud, and it was close by. I got up and went downstairs, where my landlord and landlady were already up, and donning their outdoor coats over their night attire. I quickly joined them, and we sallied forth rather gingerly into the night, to search for the origin of the noise that had disturbed us. It didn't take us long. Draped over the roof of a house just a few doors along was a parachute, with its harness dangling down over the front of the house – but there was no one dangling from it. The Home Guard were soon on the scene, and with great excitement began searching for the missing airman. One of them was scouring every nook and cranny, including some allotments nearby, and as he peered hither and thither he was calling out excitedly, in a broad Somerset accent, 'Now cum on –where's zum to? where's zum to?.' The German airman could have been forgiven if he had thought he was being addressed in broken German.

They found him. He turned out to be little more than a schoolboy, perhaps 19 years of age, and he looked terrified. I think he feared he was going to be summarily executed.

What quarrel had he with us English? The thought immediately brought to mind the film which had been largely responsible for my becoming a pacifist. It was called 'The Road Back', and I had seen it at least twice, some years before the war. It told the story of a German soldier returning home immediately after the 1914 – 1918 war. He went to find his girl friend, only to come upon her with another man. With hardly a second thought he drew his pistol and shot the man. He was tried, and convicted of murder, and it was his plea in his own defence that affected me so much – so much, in fact, that it changed the course of my life. 'For four years,' he said, ' I have been required to kill at sight men with whom I had no personal quarrel, and I was praised and indeed rewarded for doing so. Now, when I kill a man who has taken away from me someone I loved dearly, you call it murder.'

It wasn't that I condoned the shooting, it was the double standards employed in justifying war that changed my thinking so decisively. I wonder how many more who saw that film were similarly affected by it?

My post at the Bristol school was for two terms only, Autumn 1940, and Spring 1941, and nearing the end of the Autumn term I spotted a post advertised in the Times

Educational Supplement which was at Bradford Grammar School, and, what was more important, it was 'for the duration'. I applied for it, and was offered it, on condition that I could start at the beginning of the Spring term. I went to the headmaster of the Bristol School and explained my dilemma. He may have had little understanding of discipline, but he was very understanding of my problem, and released me from serving the second term.

Incidentally, I had never been further north than Birmingham before I applied for the post in Bradford, and I was in for a shock. When I came out of the station it was to see building after building literally black – black with a century's soot. I honestly couldn't believe my eyes, and it took some time before I could throw off the depressed feelings the sight engendered. It wasn't just the black buildings, it was the frequent sight of a lone wind-sculpted tree on the sky-line, which had long since discovered that its best chance of survival was to 'go with the wind', so to say. But for someone who had feasted his eyes on the lushness of the Hertfordshire countryside, it was indeed a bleak spectacle.

Once again I was fortunate: the headmaster at the new school was one Richard Graham, another Quaker. Bradford Grammar School was largely fee-paying, and the lowest stream (out of six) consisted mostly of the distinctly backward sons of wealthy mill owners. The headmaster gave me a list of possible digs, and the beginning of the Spring term found me ensconced in very comfortable and homely lodgings in Shipley, a mile or two from the school. My new landlord and landlady were a Mr and Mrs Tilling, Yorkshire to the core, very friendly, jack-blunt, and honest as the day. Like my former landlady Mrs Watterson, Mrs Tilling was an excellent cook, and very proud of it, too. I shall always remember the scorn with which she spoke of those who made custard without eggs. After all, it was wartime, and I never did learn where she obtained what was obviously a plentiful supply of eggs. (No - they didn't keep chickens!)

Now that I was in a job which was 'for the duration', Mary

and I felt able to get married – at last! – for, by the time we were married, we would have been engaged for almost three years. It meant, of course, finding somewhere for us to live, and once again fortune favoured me. A mile or so beyond Shipley was the town of Bingley, that might have been described as small and undistinguished but for the fact that above it on one side was the village of Haworth, the home of the Bronte family. On the other side, the road climbed to the little village of Gilstead, which was to be our future home, but which was already the home of Fred Hoyle, the cosmologist famous for having proposed the theory of 'continuous creation', as an alternative to the 'Big Bang'. All I can remember of Fred Hoyle's theory was that throughout the expanding universe it would need only one atom of hydrogen to be spontaneously created per year in a volume equal to that of St Paul's Cathedral to make up the necessary matter. His theory lost out in the long run, but I can still remember that for me, as a physicist, his house seemed to have a certain aura about it which I felt aware of, each time I passed it. After all, you can't work on a bigger theory than one concerning the Universe and its creation.

The house I was lucky enough to find (to rent, of course) was Number 5 in a small crescent called 'Gilstead Drive'. Though it couldn't have been known at the time, it was already the home of a two-year-old destined to become famous - one Harvey Smith, whom I suspect became much better known to the general public than Fred Hoyle. Yes! – as the future famous show jumper, I remember his mother telling us that 'he had a passion for horses', and refused to sleep in his cot because he wanted 'to sleep on the floor, like the horses'...

We were married on Saturday, August 2nd, 1941, at 8 o'clock in the morning. The main reason for that was an unpleasant one. By then I was very well known in the little town of Stevenage for my pacifism, and, in consequence, had come up against considerable hostility at times. So much so that I had got it into my head that someone might try to ruin the service – something that I thought would be less likely at such a time of day. The early wedding also meant that we could get away to a good start on our long journey to Bradford: there was to be no honeymoon in wartime. There was something else, which I have regretted ever since – we didn't even attend the Reception. Such were the times we lived in.

In many ways Gilstead was an idyllic spot in which to begin married life. It was on the very edge of Bingley Moor, and only five minutes' walk from our house was Shipley Glen, a local beauty spot. It was good to be in a village and a small crescent of only about ten houses: like that, it was in itself a small community in which we got to know each other very well.

The school was right in the centre of the City, and the building had certainly seen better days. Actually, every day, commuting from Bingley to Bradford on the bus, I passed its beautiful, brand new, stone-mullioned replacement. It was just ready for occupation when war broke out, and the army had taken it over. Among other things, one wonders what army boots did to parquet floors. Its take-over by the army was a very sore point among the staff of the school.

I didn't learn it until later, but the fact was that I was taking the place of a teacher whose classroom discipline was so bad that it had reached a climax when boys lit bonfires in their desks... The worst-behaved (but highly intelligent) form was Transitus Upper, and I took them within a day or so of starting the job. I was lucky. When trouble threatened to break out, by sheer good fortune I picked on the ringleader (though I didn't know it at the time), and packed him off to the Head Master. I think the rest of the form thought I had some sort of sixth sense, and from then on I had no trouble at all with them. In fact, I enjoyed very much teaching them, as they were *extremely* bright. I well remember as an end-of-term treat doing some very complex mathematical puzzles with them. They were brilliant, and very much appreciated being 'stretched' intellectually. And I very much enjoyed stretching them.

I have some recollections of my colleagues. There was a certain 'Mr Hardy', who taught science to the lower forms. He would have regarded himself as a 'gentleman' teacher -

someone who had chosen to teach as something to do, rather than as something he had to do for his daily bread. His classes were little short of a riot at times. Then there was a New Zealander – I can't remember his name – who taught Latin. You could hear his classes yards and yards away down the corridor. This was not through any lack of discipline: it was his teaching method – highly successful too it was. He would have them singing the conjugation of verbs to rollicking tunes. It went down very well with the boys, and there's no doubt the conjugations went down very well, too.

Almost all the masters treated me sympathetically. There was one exception – the Second Master. His name was Glassey, and without the 'e' it would be a suitable word with which to describe the kind of look he frequently gave me – he avoided actually speaking to me whenever possible. One particular memory of him comes back to me – of him carrying out one of the tasks that fell to him as Second Master: that of constructing the Timetable for the whole school. It was a large school, little short of a thousand pupils, divided into 35 forms. The job was consequently a very complex one, which he carried out with the aid of labelled flags attached to pins which he moved hither and thither on a large cork board. As with myriads of other mundane but complex jobs, what light work the computer has made of it now! (And that goes for what I'm doing at this moment – typing, of course!)

So much for the staff – but what of the pupils? In general, I must say I liked those Yorkshire lads, and I liked teaching them. Once they realized that you weren't going to stand any nonsense, and that you meant business when it came to teaching, they knuckled under and responded wholeheartedly. Of course, a school as large and prestigious as Bradford Grammar School was bound to have its quota of famous sons. There was a time when I knew quite a few of their names, but at the moment I can recall only one – Dennis Healey, the ex-Chancellor of the Exchequer of the Labour Party, who had, however, left the School some years before my arrival.

There was one pupil I taught, though, whom I became convinced was going to make a name for himself – and I was right.

He first came to my attention when I confiscated a sheet of paper he was surreptitiously handing to his neighbour. It was entitled 'Wisdom Sold Cheap by Donald Rooum - All Proceeds to go to the Cancer Fund'. I thought to give just a few samples of what followed, but found it impossible to choose: the document had a shape and form of its own, which brooked no omissions. So here it is, *in toto.*

Daft stories	1d
True Stories	1d
How to do Maths (per sum) (not perfect)	1d
How to do German (per sentence)	1/2d
How to do English (per sentence)	1d
Teacher's writing read (per sentence)	1d
Teacher's advice, in simple English	1/2d
Spelling any word	!/2d
LOAN OF ANY BOOK	
In dinner hour	1/2d
On night before handing in	1d
On any other night	1/2d
Look at pocket book	1d
EXCUSES	
To teachers (small)	1/2d
To teachers (big)	1d
To prefects)	1/2d
HOW TO CONDUCT ONESELF	
For Headmaster	1½d
In detention	1/2d
For Head Boy	1d
At Prefects' Meeting	½d

Lesson in any period	(per dozen)	3d
How small things work		1d
Miscellaneous Wisdom		1d

I think it was wrong to say that I confiscated that sheet of paper – it was so pricelessly witty that I would have found it very difficult to return it – and I have kept it by me ever since.

Donald Rooum distinguished himself in another context. Religious Knowledge was not a popular subject among most of the staff, so I volunteered to teach it, and the Fifth Form I was allocated included, yes - Donald Rooum. To make life a little more exciting, at the end of the Autumn term I used to set an essay on a topic we had discussed, and on this particular occasion it was Paul's voyage and shipwreck. Donald Rooum turned in 16 foolscap pages!! - much of it completely hilarious. His prize was a copy of Dorothy L Sayers' 'The Man Born to be King' – a wonderful series of plays for radio, written and broadcast during the war.

About 50 years on (around 1990) I came across a reference to one 'Donald Rooum', from which it was plain that I had been right about him – he had become quite well-known, and was into journalism. Later (just a few years ago), I discovered that he had his own Website, and, as you might expect, very impressive it was, too. He was not only a journalist, but an accomplished political cartoonist as well. His name, entered in 'Google', produced dozens of references, many of them showing him to be active in the anarchist cause, but many others clearly demonstrating his concern over social issues, such as child abuse. It is a fascinating experience to have had a teacher/pupil relationship with someone who had subsequently developed into a notable personality, and, indeed, to have sensed that he would do so. I have promised myself that I will send him a copy of that essay of his. Will he own, or disown it? I wonder.

There was one other pupil at the school whom we got to know very well, and, indeed, became a personal friend of ours. He was on the Classics side, so I did not have him as a pupil at the school, but he was in my Bible Class at Gilstead Methodist Chapel. His name was Alan Keighley, and he lived close to us in Gilstead village. His father was a skilled cabinet maker, as my father was, and his mother a typical skilled Yorkshire housewife and cook. Tuesdays was baking day in the village, and how well I remember calling on the Keighleys on a Tuesday, to be greeted by that wonderful smell of baking bread.

Alan subsequently became a Methodist minister, and our paths seemed to cross quite frequently – not only metaphorically, but geographically: for example, when I was working at the National Physical Laboratory in Teddington in the 1960s, he was the minister at a Methodist Church in neighbouring Surbiton. Tragically, he died soon after, of a brain tumour.

So much for Bradford Grammar School, and its ramifications in my life. I cannot end this Chapter without saying a little about one more friend we made in Yorkshire.

On the way up to Gilstead, and on the opposite side of the road to the hospital, was Bingley Teachers' Training College. As far as I can recollect, it was Mrs Keighley who introduced us to one of the lecturers there – one Louisa Roberts. She was a Quaker, and a person with many deep concerns. For example, she was a prison visitor (on behalf of the Quakers) at Strangeways Prison in Manchester. It wasn't long before we had a house group meeting regularly in our home, of which Louisa was the unofficial leader. Indeed, the meetings always began with what amounted to a brief Quaker Meeting, before going on to discuss some issue of the day. The group, meeting in our own home, became very much part of our life in Yorkshire, and I think it would be true to say that it was our friendship with Louisa, more than anything else, which led eventually to our joining the Society of Friends, nearly ten years later. And when the time came for us to leave our little house and home in Gilstead, it was the deep fellowship within the Group (just like the Stevenage Group, when we had gone north), that we would miss so sorely.

Chapter 7

As we have seen, it was on August 2nd, 1941 that Mary and I had settled down to married life in Yorkshire, for what we expected to be the duration of the war. And but for the war (and it's a very big 'but'), it was a life that could have been described as quite idyllic, situated as we were within a stone's throw of lovely moorland walks, and the Dales a mere bus ride away.

As things turned out, we were soon into married life in earnest: our first child, John, quickly on the way, and born six weeks early, on May 11th, 1942. One day when I arrived home from school, to my consternation both Mary and the baby were missing. One of those friendly neighbours had been looking out for me, however, to tell me that Mary had sent for the doctor to look at what would normally have been John's navel. He had taken one look at it and said that John must go to hospital at once – he had an umbilical rupture that was in grave danger of bursting. Those were the days when GPs were, indeed, *general* practitioners: the doctor did the operation himself at the cottage hospital, which was situated almost next door to his surgery, and on the road up to Gilstead. I went straight back down the hill again to the hospital, to find Mary keeping vigil beside John, not quite due to be born still, and looking truly awful. I remember only too well the words which came into my mind just then - though, needless to say, they went unspoken: 'We're going to lose him, we're going to lose him.' John was of tougher material than that, though, and made what is called an 'uneventful' recovery.

About a week before his operation, I had celebrated my 29th birthday, and Mary had given me a book as a present. It was the biography of Marie Curie by her daughter Eve. From Mary's point of view it just seemed a suitable book to give to a physicist, bearing in mind its subject matter. How was she to know the dramatic outcome it would have? The fact is that, almost four years into a war with no end in sight, I was beginning to get restive in my teaching job, especially as it had involved the move north, away from London and the south-east, where the action was at its fiercest: the move itself was, of course, incidental to the job, but I had always felt that it could be misinterpreted by those who knew that I was a CO. But there was a deeper layer - a much deeper layer - to my feelings on war in its widest sense: I think my mother's extreme fear of cancer had determined very early on in my life that, one way or another, cancer would become the battleground on which my personal war would be fought. And as I read Marie Curie's biography - of her Herculean struggle to isolate a miniscule amount of radium from 7 tons of pitchblende ore – it came to me that war as such was not the issue, *it was the nature of the weapons, and against what or whom it was being fought.* My war would be against cancer, and it would be fought in a laboratory. That way, to put it vulgarly, I would be killing two birds with one stone: I would be facing up to the legacy of fear which my mother had passed on to me, and at the same time I would, in effect, be making a statement about the kind of war into which I was prepared to throw my whole weight – not just 'for the duration', but for the rest of my working days. I told Mary what she had set in motion, and she was wholeheartedly with me in what I proposed to do about it. And that was to write to the Tribunal which had stipulated that I should teach, not to ask their *permission* to give up teaching, but simply to tell them of my intention to do so. I expected a tussle with them, but, when it came to it, I heard nothing at all from them.

Strangely enough, I was able straightaway to take practical steps towards the realization of my plans. The school timetable was arranged around Tuesday, Thursday, and Saturday afternoons as games afternoons (with ordinary school on Saturday mornings). What is more to the point, there was the Bradford Royal Infirmary, situated on our side of the City, and it had a Radiotherapy Department. So - I wrote to the Physics Department, giving my background, and asking whether I could serve as an unpaid assistant on Tuesday and Thursday afternoons: I would be an extra pair of hands, and at the same time serving some sort of apprenticeship which would stand me in good stead when it came to applying for a full-time post in Hospital Physics. It was a somewhat unorthodox proposal, but it was accepted enthusiastically. They had just one qualified physicist, a Miss Jones, who turned out to be a good teacher, and her enthusiasm for the work matched mine. We got on very well, and I attended the Infirmary for the better part of a year, learning among other things how to measure X-ray doses, and to calculate the gamma-ray doses delivered by arrays of radium needles, when applied to cancerous areas of the body. Incidentally, I was becoming conditioned to what it would be like to work in a large teaching hospital - not that the following incident should be regarded as typical of what went on – quite the reverse, in fact!

I arrived one afternoon to find that a medical student had set himself up in a corner of the laboratory to perform his allotted task of dissecting a human leg. It was what one might call a 'vintage' leg, with green mould growing on it! Miss Jones protested mildly, but when Sister came in she was livid. It was, of course, a dangerous thing to have done, and it could be said that the student beat a more than hasty retreat, complete with the leg...

I myself was guilty of a mistake which could have had serious consequences. I had a supply of radium needles, and was constructing one of those radium 'plaques', the array of radium needles being contained between two areas of sticking plaster. I was working over the top of a 4-inch wall made of lead, which protected most of one's body from the gamma-rays. (Hands, and arms, and head had to take pot-luck, so to say.) Somehow or other I inadvertently dropped a radium needle on the floor, unaware that I had done so. However, the beady eyes of Sister soon lighted upon it, and I got a ticking-off nearly as severe as the medical student. Incidentally, because of the wartime conditions, the main radium stock was kept down a deep bore-hole, many feet below the surface. This would have prevented a catastrophic scattering of naked radium in the event of the hospital receiving a direct hit from a bomb.

During this 'apprenticeship', my brain had already started

Plate 8 First version of Linear Radium Source Dose Calculator

to get busy on some ideas I was having about the computation of gamma-ray doses delivered by those implanted arrays of radium needles. The extraordinary thing was that these ideas came under the category of what, nowadays, would be called *analogue computing,* a term quite unknown to me, and, I suspect, to most other people at that time. Be that as it may, it wasn't long before I was assembling what in fact amounted to an analogue computer, *literally* on the kitchen table (I had no access to an engineer, so the components were all of wood, and fabricated for me by my craftsman father - Plate 8). In fact, I felt some humble kinship with Marie Curie, who had so often to work under makeshift conditions. It was a start, and would certainly make a good talking point when I applied for that full-time post which I had so keenly in my sights.

That turned out to be at The London Hospital in Whitechapel Road, in the East End, where I commenced work on April 5th, 1944. I was to work under Dr John Read, on behalf of Dr Frank Ellis, the chief Radiotherapist. Both of these already had international reputations, so I was off to a good start, and Dr Read was soon encouraging me to work on my dose-computing ideas. The upshot was that as early as October, 1944, I proudly published my first scientific paper – it was called 'Linear Radium Source Dose Calculators', and was the first of a total of 43 papers, published over a span of 30 years or so. I had, indeed, found my life's work.

It had, of course, necessitated moving south again, and, in fact, for the next two years we lived in Stevenage with my parents, and I returned to daily commuting to London. Whitechapel was easier to get to than North Wembley, however, though it wasn't long before the V1 and V2 raids began, which added to the stresses and strains of travel. One didn't necessarily go by the shortest route, so much as one that ensured that as much as possible of the journey would be underground. The so-called 'buzz bombs' weren't too bad – at least you could hear them coming, and even watch them, gauging where they were going to land. But the V2 rockets were another matter altogether – they hurtled down at supersonic speeds, and in consequence you only heard them after they had landed, and, of course, that meant that you were safe - from that one, at any rate. The London Hospital, situated in the heart of the East End, had to cater for the casualties involved, should one of them land in the neighbourhood. I remember once taking a shortcut through a hospital building I was not familiar with, only to find that the corridor in it had been lined with hastily constructed metal racks, stacked with the bodies of victims of the latest raid. There must have been forty or fifty of them.

Mary was pregnant again when we made the move south, and our second son, Roger, was born in the hospital of the neighbouring town of Hitchin, on August 11th, 1944 – four months after the move. In 1946, though, we were on the move again. We had bought one of the first new houses to be built after the war – it was in South Ruislip, west London, where my sister Vera and her husband Gilbert lived. It cost us a mere £1200 pounds! - but that would probably correspond to something approaching £100,000 nowadays. It almost goes without saying that it was 'jerry-built'. No such thing as grooved and tongued floorboards – just plain planks, and how well I remember rugs actually flapping up and down on the floor when the wind blew! Still, the house served its purpose - bridging the three-year gap between 1946 and 1949, when we purchased the house that we were to settle down in for the next 29 years. It was on the western outskirts of Guildford, 28 miles from Waterloo, on what was then called the Southern line. Yet again I was commuting to London, but it was by fast train that took little over half an hour for the journey; the worst bit was getting from Waterloo to Whitechapel, on a grossly overcrowded underground link between Waterloo and Charing Cross (commonly known as 'The Drain'), and then on to Whitechapel via the District Line.

Against this backdrop of developments in our family life, there had been - in 1946 - an important development in my work. Dr Read left to join the world-renowned research team at Mount Vernon Hospital, in Northwood, and I was offered his post as head of the Physics Department.

It was a considerable honour, and a very considerable increase in my responsibilities too, but it was a very fertile time for me: by December 1946, I had published 7 papers on the measurement of X-ray and gamma-ray dose and dose distributions. These described entirely new measurement techniques, including one enabling the direct comparison of two ionization currents (it even came to be known as 'The Kemp Cpmparator'!), which had made possible the development of an automatic X-ray dose plotter. This could plot the entire dose distribution in an X-ray (or gamma-ray) beam in about an hour and a half, which hitherto, 'manually', had taken a *day* and a half. (In the early 1950s I submitted a thesis based on this work to London University, and was awarded a Ph D. on the strength of it.) I had also broken new ground by venturing into the X-ray Diagnostic (Radiography) Department, demonstrating that there was much work there for the physicist, in clarifying and optimizing radiographic techniques.

The mere mention of the automatic dose plotter was bound to evoke one of my most vivid memories – namely, of the moment when I switched it on for the very first time, having 'told' it to plot a particular isodose curve (i.e. line of equal dose) in an X-ray beam. I was on my own, having worked on into the evening to reach this dramatic moment, and as I watched, it no longer seemed to be a mere piece of equipment, as it embarked on its allotted task: it seemed to come alive – a creature that had listened to what I wanted it to do, and had set about it with a will. As I say, I was on my own, but suddenly felt a desperate need of someone to share the moment with me. I went out into the corridor, hoping to find a medic, or some such, who might also be working late, but there was no-one except stolid East-enders, seated on the benches, waiting to see patients (it was Visiting Hour). I turned back into the laboratory, to find my faithful servant still busy with its task; I was feeling very frustrated at not being able to share my special moment with anyone.

Alas, the feverish effort behind all this, and the new and

heavy responsibilities involved, proved, for the time being, too much for me. In just over two years I had moved from being a junior master at a Yorkshire Grammar School to Head of Physics at a very large London hospital.

So it was that, in 1947, I had a breakdown.

I was supported through this critical period by our family doctor, Willoughby Clark, who was also practising as a Jungian psychoanalyst. He helped me clear a lot of junk out of the attics and cellars of my mind, and I returned to work older and wiser for the experience. Willoughby later moved to Harley Street, to practise psychoanalysis full-time. By then he had become a close personal friend, and well do I remember calling on him for a cup of tea whenever I needed to visit the library of The Institute of Radiology, in Welbeck Street. He was a very wise and kindly man, but, alas, he had his own problems: he developed a bipolar condition, which finally ended his Harley Street days. In his hyperactive times he would go days and nights without sleep, turning out a host of articles, essays, and poems, some of them almost Shakespearean in nature and quality. At one such time he threw a lavish party, and I remember all too well a banner on his sitting-room wall, putting an entirely new slant on Christ's words, 'My burden is Light.'

Chapter 8

Our move to Guildford in March 1949 combined with a number of other events to make 1949 very much a year of new beginnings for us. There was, of course, the birth of Rosemary Joy on January 18th - I remember so clearly that second Christian name coming to me as I cycled over to Harrow to convey the happy news to some friends of ours. The third 'new beginning' was not quite so obviously such, but was, in its own way, just as significant — namely the fact that, a year before, I had emerged from a deep and searching Jungian analysis which had given me many insights into the nature of the forces which had driven me so hard, and had resulted in that veritable torrent of 7 papers published in under three years. The fact is that by 1949 I had made a new beginning in my approach to my work — it was more balanced and level-headed, and no longer had total possession of me after the manner of religious zeal. There was still a good head of steam in the boiler, but the psychoanalysis had installed a safety valve.

As will be seen, the next few years were largely 'harvest years', when the instrumentation developed during those first three hectic years would bear a rich crop of what came to be standard dosage data for both radium and X-ray sources and widely adopted as such, but my research work itself took a new turn — a study of the fundamental units of radiation dose and their realization, which had an extraordinary outcome.

It was in 1949, at the Scientific Instruments Exhibition under the auspices of the Institute of Physics, that I exhibited the final, and very sophisticated version of the Linear Radium Source Dose Calculator - which had, by then, certainly taken on an appearance worthy of the term 'analogue computer'! (Plate 9). By an extraordinary coincidence I found myself next to an exhibitor by the name of Colin Cherry, with whom I had worked ten years before in the television laboratory of the General Electric Company in North Wembley. By now he was Dr Colin Cherry, Professor of Cybernetics at Imperial College, and indeed it was he who informed me, to my great surprise, that I was

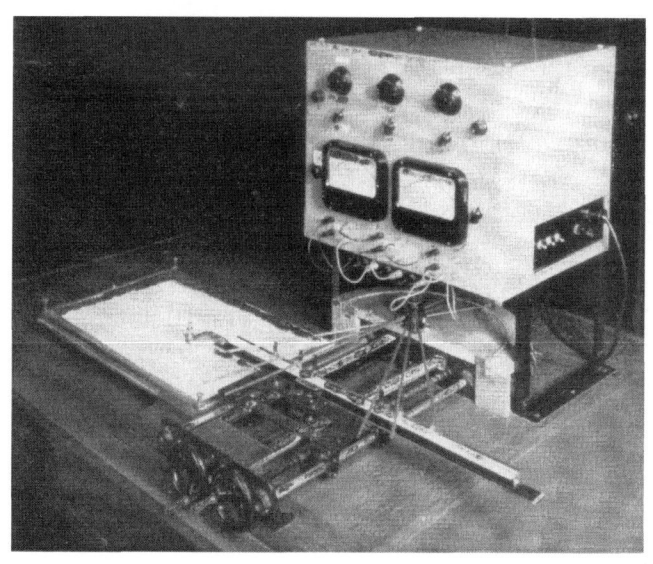

Plate 9 Final version of Linear Radium Source Dose Calculator

working in the field of analogue computing! My exhibit enabled one to plot the lines of equal dose ('isodose curves') round assemblages of up to three different radium needles 'in tandem'. Such arrays were in common use in gynaecological treatments, and my analogue computer enabled the changes to be rung on these arrays with great ease.

The automatic X-ray dose plotter performed a similar function (but by direct measurement) for all the various X-ray beams (of different widths and penetrating power) used in treatments, plotting isodose curves, in effect, within the body of the patient, as percentages of the dose received at the surface. (No, the probe did not explore within the actual patient! For most purposes it is sufficiently accurate to represent the patient by a tank of water(!) called a 'water phantom', through which the probe moved, to find those lines of equal dose.)

By 1952 these comprehensive surveys of gamma-ray and X-ray dose distributions had been completed and were in wide use; and, as already foreshadowed, I had moved into the field of the basic units of radiation dose, and their realization at the international level, which was - in England - the responsibility of the National Physical Laboratory in Teddington, Middlesex, and in the United States the Bureau of Standards, in Washington. I had come to suspect that all was not well with them – a suspicion which I could barely entertain. Nevertheless, our work at The London Hospital, which to begin with concentrated on the British standard, led us to the inexorable but barely believable conclusion that it was in error by about 3%. The two national standards had been compared in the 1930's, and been found to agree to within 1%, regarded at the time as very satisfactory. The possibility that they might *both* be wrong, yet agree to within 1%, was not even considered at the time – after all, why should it have been? But that is precisely the conclusion to which our work was pointing. With considerable trepidation we turned our attention to the American standard, and in due course the case was proved: it was about 2% in error, which made possible the agreement to within about 1% with the

Plate 10 Apparatus used to measure the Errors in International X-ray Dose

British standard.

Without going into details, the apparatus we used for this momentous work is shown in Plate 10. For the record, I will mention just one thing, however, concerning the equipment. The so-called 'Great Smog of London' occurred one weekend during the course of this research, and a bank of high-value resistors - at a considerable voltage by virtue of being connected to the stack of batteries seen at the bottom of Plate 10 - took on the appearance of *hairy caterpillars* as the result of the smog particles attracted to them by the high voltage. And these (largely carbon) particles *short-circuited* the stack of batteries, reducing their 2500 volts to zero, overnight. And to think – we were *breathing* that air...

Well now - to misquote Oscar Wilde: to assert that one international standard is in error is to stick one's neck out, but to assert that two are in error, yet in agreement, is to put it on the block!

That is precisely what I did in a paper read at the International Conference on Radiology held in Copenhagen in 1954, but I had made doubly sure of my facts, and survived! Dr L S Taylor, who was in charge of the American standard, was present at the time, and came out with a comment on the work we had done that I have never forgotten. It was spoken in a broad American accent. 'Well,' he said, 'it only goes ter show that if yer sittin' on the fence, you ain't necessarily sittin' pretty!'

A paper describing the work, and read before the British Institute of Radiology - in effect as a dress rehearsal of the paper read in Copenhagen - was deemed 'the most meritorious contribution delivered before the Institute in the year 1953 – 1954', and was given the 'Roentgen Award' for that year. It was the climax of my first ten years in radiation physics.

The year 1954 turned out to be a milestone in my career for an entirely different reason. The so-called 'atomic pile', or nuclear reactor, as it came to be called, was developed during the war specifically to produce material from which atomic bombs could be manufactured, but after the war they began to be used for the production of artificial radioactive isotopes in much larger quantities than the amounts of radium available, so that new machines could be constructed producing powerful beams of penetrating gamma-rays equivalent to X-rays generated at a million volts or more. In particular, by the early 1950's, radioactive cobalt was being used in such machines, protected by a massive sphere of lead, except for the aperture through which the gamma-ray beam emerged. At the time, and because of the spherical shape, they were known as 'cobalt 'bombs'. But more of that a little later!

In 1954 The London Hospital decided to purchase such a machine, and the chief radiotherapist, Dr Boden, and I were commissioned to go on a fact-finding journey across Canada, and back across the States, before the final decision was taken as to which machine to purchase – they were, of course, very expensive. You might say that Dr Boden and I went on an atomic pilgrimage. And, for sure, 'Thereby hangs a tale' as they say, and it is certainly deserves to be told in some detail. So, here goes!

It must be pointed out straightaway that it was in the reign of McCarthy – when the fear of Communism was so bad in the States that one was almost assumed to be a Communist unless you could prove otherwise. Anyway, it meant that we had to apply for visas for our visit to the States, and at the same time supply a bank letter stating that funds for our support whilst in the U S A were available, and a letter from the House Governor of the Hospital, stating the purpose of our journey.

My first mistake was to go to the American Embassy itself, instead of No. 25 on the opposite side of the Square. Beating a hasty retreat from the diplomatic circles into which I had inadvertently strayed, I handed to the young lady at the reception desk at No. 25 my passport (and 'three identical photographs of myself, two inches square, full view, without hat, taken against a light background, and printed on thin paper'...), together with the bank letter, and the one from the House Governor. 'Would you mind sitting over there, please - in the front row,' the young lady said, with a barely perceptible drawl. I obeyed, and found myself with others, just outside a door over which an illuminated sign 'Next Please' lit up from time to time, accompanied by the sound of a buzzer – for all the world like a doctor's surgery. The sign lit up, to the accompaniment of an imperious buzz. It was my turn, and by then it was with some apprehension that I went through the door and sat down at a large desk, opposite a dark, somewhat sombre but business-like young man. I handed him my papers, feeling like a little boy who had just given his school report to his father. A full minute passed. 'Are you a physicist?' The question was put slowly and deliberately. 'Yes.' The screw was given another half-turn. 'Are you an *atomic* physicist?' I tried to be jocular. 'Not really,' I said. He pressed on, regardless. 'Are you interested in atomic energy?' 'I'm interested in the medical applications of atomic energy. In fact, I hope to visit Oak Ridge during the course of my trip.' And I believe that in an unguarded moment I may have added something about needing to find out about the production of radioactive sources for use in cobalt bombs. The word appeared to sting, like a joke in bad taste, until the young man suddenly realized I meant it. Without another word, then, he rose and disappeared into an inner sanctum. He came back, looking very solemn. Sentence was passed. 'We're gonna route you specially – treat you differently from our usual run of customer, and that's gonna take a little longer than usual. When do you wanna sail?' (It was in the days when the 'Queen Mary' plied regularly to and fro between Southampton and New York.) 'August the Twenty-first,' I said, blandly. 'Struth, you're cutting it a bit fine.' He sounded serious. 'We'll do our best, of course,' he added, 'but in the meantime you'd better fill this up right-away.'

He handed me a foolscap sheet spattered with questions on both sides. My heart sank as I scrutinized it... 'names in full, including any aliases... age... birth date... parents' names... and address... wife's names... including maiden name... every organization etc that you have ever belonged to...' I looked up apprehensively. 'I am a Quaker. Is that the sort of thing you want to know?' 'I don't mind if you're a freemason. You know the sort of thing we are after.' I detected an element of discomfort in his reaction, as though he sensed what the form must look like to the average Englishman. The young lady at the desk came in. 'Can you do one of your one-a-minute efforts? There's an 'ellova queue outside.'

I felt I was becoming a trouble spot, and returned to scrutinizing the form. 'Name and address of an American referee... name and address of an English referee... purpose of visit... arrangements for travel... ' I crept away to a corner of the reception hall, where there was a writing table, and started to reconstruct my past. 'Names, ages, and addresses of all children, brothers, and sisters... Have you ever been to East Germany? If so, why?...' Then came a request to state everywhere I had lived since 1939, including street, and house numbers. The questions ended on a solemn note: 'Have you ever been arrested... convicted... deported?'

There was still an affidavit to sign, to the effect that I was not, and never had been a Communist. I was tempted to write, 'No, and I have never even wanted to be,' but I desisted. After all, I was in enough trouble already, without making things any worse for myself.

My passport was handed back to me, together with all three identical photographs (for some odd reason this felt like a personal slight), together with the rest of my documents, and I was told that I would be hearing from the Embassy 'in due course'. 'But don't you want my fingerprints?' 'No, thank you — not this morning.' I was disappointed. I realized I had been nurturing a secret delight at the very thought of being fingerprinted, but he was adamant: until my story had been substantiated, they weren't going to waste time taking my fingerprints.

I trooped out of the room. From the dim recesses of my mind I recalled that the travel agents had said that I must have a valid certificate of vaccination. A sudden doubt assailed me.

Vaccination against what? I had assumed that it was something as comparatively harmless as smallpox — but what if it were Communism? Was that what I was required to be vaccinated against? Ah, me! — if only they had taken my fingerprints...

The next morning a physicist from St Thomas's Hospital who was to share the greater part of my journey, on an identical mission, presented himself at the Embassy, having been warned by me of the complications that might ensue if he delayed any longer.

The dark, but not unfriendly gentleman who had interviewed me the day before looked up as he entered, and asked him what the time was. My friend said, '12 30.' 'Lunchtime,' said the dark gentleman, with obvious interest. 'Yes,' said my friend. 'I suppose I am your last customer before lunch.' 'Sure!' A quick glance down the papers, the odd question or two, and my friend was out again in the street - with his visa! Admittedly, his papers hadn't actually mentioned Cobalt 60 - merely referring vaguely to 'teletherapy units'. But there, ignorance being bliss, the junior consul presumably enjoyed his lunch as usual that day. Even those engaged in potential witch hunts needs must eat...

Three weeks later, I was back again at the Embassy to enquire how my visa was getting along. My papers were in Washington, I was told, and I must await developments. I pointed out that it was little more than a fortnight, now, to my sailing date. 'What are the prospects?' 'Fair - fair,' he hedged. 'What are the chances of obtaining a transit visa to enable me to land in New York and pass more or less straight through to Ottawa? - perhaps my full visa could catch up with me in Canada before my real business in the U.S. began.' He would see what he could do along those lines, he said. In the meantime, was my hospital prepared to pay for a telegram to Washington? I said I thought so, in the circumstances. I felt sorry for him. He seemed to be doing his best to get me out of an awkward situation. 'Some of the guys around here would just let things take their course. The trouble with me is, I'm soft,' he said.

A few days later he telephoned me to say that the transit visa was out of the question but that he had sent a cable to Washington, giving as many good reasons as he could think of, in support of the earliest possible reply.

Again I had the impression that a considerable effort was being made on my behalf, but the impression was to fade in the course of the exasperating events which followed in the next fortnight, and which led inexorably to the cancellation of my boat passage. A week before the sailing date, an official at the Home Office contacted the Embassy on my behalf regarding the delay, and two days later the Secretary of my hospital asked if another cable could be sent to Washington, in spite of the fact that no reply had yet been received to the one sent ten days earlier; and, furthermore, he expressed willingness to omit from my itinerary a visit to the U.S. Atomic Energy Establishment at Oak Ridge. This would leave us with a number of visits to hospitals, and to manufacturers of equipment in the medical field, in which McCarthy himself could hardly have found cause for objection. He was told that such a step would not accelerate matters, and that there was no point in sending another cable at the moment. We began to get some idea of the juggernaut we were up against - an enormous army of civil servants 3000 miles away in Washington, completely impervious to the sense of urgency that (in those days) normal human beings associated with telegrams. Seventy-two hours before we were due to sail, I rang the Embassy yet again. I asked the Junior Consul if he remembered me. 'Sure! I think about you every day and every night.' I said that was nice of him, but it didn't get me my visa. By this time I was fast becoming both a desperate, and despairing man. Was there any news at all? No, he said, there was still no reply to our cable. I repeated our request for a second cable to be sent. His reply was that he couldn't order a second one without the Consul's permission. 'Why?' 'Well,' he drawled, 'communications tend to get blocked, you know.' 'Whose communications?' I queried. 'What about mine? Aren't they being blocked in a big way?' 'Sure, they are,' he agreed. Then, suavely, 'You'd better have a word with the Consul. He'll be here in an hour from now.' And he promised to place my file on the Consul's desk. I rang off, and waited. On the stroke of the hour, I rang again, and was put through to the Consul immediately. I explained the position, and asked why so little had been achieved in the five weeks that had elapsed. 'These things take time.' 'What things?' This was no idle question. 'I should be happier of I hadn't discovered today that you haven't even taken up my English references'. It was not usual to do so, he said. Then what on earth was taking the time? I asked, conscious once more of the vast bureaucratic machine I was up against. 'Look at it my way,' I said. 'I want to visit your country on what is a purely medical mission. We are prepared to omit Oak Ridge from our itinerary. That leaves hospitals and medical organizations only. Why - why - *why* all the fuss, after five whole weeks?' He was silent on this issue, but agreed to send a cable, provided, he added, that I was prepared to leave a deposit of four pounds...

I went straight off and handed it to an extremely discourteous young lady, who behaved like a school ma'am dealing with a factious young boy. It was at this juncture, I confess, that I finally lost my patience: in a Quakerly turn of phrase, 'I dealt faithfully with her'.

The next day I rang twice. Nothing doing - and I fell to sorting out the complicated relationship between working hours in Washington and London, to try to decide when a reply was most likely to come.

Twenty-four hours before our sailing time, and the day began with fifteen minutes on the telephone to the Junior Consul, who was fast becoming a close acquaintance of mine. 'I am what you Americans call "all washed up", aren't I?' I said, simply. He was hurt. 'Gee! I wouldn't say that. We'll probably hear something any minute now. There's many a man saved at the gong in this game.' I didn't share his optimism. 'You must admit that from an Englishman's standpoint this whole business is well nigh incredible.' He admitted it with real embarrassment, I thought. I resurrected the idea of a transit visa to tide us over.

'No good, sir - it's against the law as it stands.' We returned to the cable. 'I haven't had my money's worth,' I said, a little peevishly. Then something came out which I had been itching to say for a long time. I reminded him that I was a Quaker, and that the last thing I would wish to do was to apply my knowledge destructively. He quite understood, he said - but it made no difference. I gave up then, and asked him to make sure he rang me the moment he heard anything. 'Believe you me,' he said, 'I've got the whole place organized on your behalf, from the roof down to the basement.' 'I wish I could believe that.' 'Gee! I wouldn't tell you a lie,' he drawled, good-humouredly. I rang off. And there it was left. The last hour or two slipped by. All that remained was the sorry duty of ringing up the travel agents to get them in turn to ring the Cunard and surrender the boat passage. How often, during those last few days, was I tempted to 'spill the beans' - to tell my friend the Junior Consul just how inefficient the system really was - the system which could select for special scrutiny one day and accept without question the next, two people of the same profession, bent on the same mission. How easy such a showdown would seem to be, and how effective! But I was convinced that it would have had only one result - the withdrawal of the visa already issued to my friend, as a 'regrettable mistake'. So my colleague from St Thomas's sailed alone, comforted by the thought that he had slipped through the net, which had been cast so clumsily. For myself, I fell to pondering again the subject of inoculation against political extremism of any brand. It seemed that we English (without being conceited about it) had an effective antidote: a grain or two of common sense, plus several grains of good humour, a dose of toleration, and the recognition that if you bottle up bad wine too tightly, you succeed only in blowing the cork, and perhaps even smashing the bottle. My visa arrived two days after he sailed.

Chapter 9

Instead of a leisurely trip across the Atlantic on the Cunard liner 'Parthia', Dr Boden and I had obtained passages on a BOAC 'Stratocruiser', to fly to Montreal via Prestwick, Reykjavik, Greenland, and the Hudson Bay. We were to depart on September 1st, ten days later than our scheduled departure on the 'Parthia'. So it was that August 21st saw me, together with my family, speeding towards Plymouth for an unexpected and very welcome week's holiday by the sea. My school friend Clifford, who was living in Plymouth at the time, had offered us the use of his house, whilst his own family were having a holiday elsewhere.

It was an odd coincidence to find myself in Plymouth, just before what seemed to me the great adventure of my first visit to the New World – for it was from Plymouth that the 'Mayflower' set sail on September 6th, 1620, with 101 passengers, who came to be known as The Pilgrim Fathers – certainly *their* first visit to the New World! The 'Mayflower' took 66 days to complete an Atlantic crossing of some 2750 miles, an average of 40 miles a day. By comparison, I was to take 18 hours to fly from London to Montreal, yet even this seems hardly noteworthy in these days when the crossing is regularly made in just a few hours.

The week passed all too quickly, and for me it was marked by a growing sense of detachment from the immediate, and the present. At such a time 'the meanest flower that blows' can evoke 'thoughts that do lie too deep for tears'. I fell to wondering what the wild flowers of the Canadian prairies, or the 'highways and byways' of the United States would be like; highways – yes – but byways? I wondered.

Back home again, and the inevitable feverish ritual of packing, unpacking, and re-packing went on, until the total weight of my luggage was just under the permitted maximum, and suddenly, then, September 1st was upon us. On a hot afternoon, the family piled into a taxi that took us off to the station in good time for my train – to miss it would have spelt disaster, since it was the last train which would get me to the Air Terminal at Victoria in time. We arrived on the platform with a little over ten minutes to spare - and it was just then that I realized I had left my raincoat at home. A trivial matter it might seem – one doesn't embark on a 15,000 mile journey without a bit of spare cash in hand. The truth was, however, that it was already late afternoon, and my last opportunity for shopping in England had gone. In the morning I would be in Canada, where a raincoat would have to be purchased with precious dollars - of which I had none too many.

Mary was ever one for resolute action in the face of a difficulty. Bidding me what we both hoped would be only a temporary farewell, and leaving the children in my care, she quit the station at high speed and hailed another taxi, which took her home and back again with a minute or so to spare - complete with my raincoat! Hasty farewells to the children (Mary was to accompany me to Heathrow) – each of whom stood for a particular and characteristic loss: no more perpetual questioning from John, no more characteristic requests for a walk from Roger, and no more silent hugs from small daughter, for six whole weeks. A final wave from the carriage window, and I began to wonder what horrendous withdrawal symptoms would be visited upon me in due course.

A reminder might be timely – that this was 1954, and flying was not the universal experience that it is today. Apart from a couple of joy rides in a Leopard Moth, each of which had lasted all of ten minutes, I had had no experience at all of flying, and it was with a mixture of apprehension and excitement that my colleague and I sat strapped into our seats whilst the crew went through what I believe is known as the 'cockpit drill'. As the engines roared to full throttle with the brakes hard on, the whole plane seemed to shake itself like a dog preparing itself for action. But when the brakes came off it was surely more like a whipped horse from which the last ounce of energy was being demanded, as it set off down the runway. Just when it seemed that the pilot must give it up, and go back to the beginning and try again, the plane became airborne. It happened so gently as to be almost imperceptible, raising serious questions in one's mind as to how it could possibly clear the buildings beyond the end of the runway. Such questions will be foreign to a younger generation that has known only jets, which are no sooner airborne, than they tilt skywards to climb at an unbelievably steep angle. In sharp contrast, our 70-ton Stratocruiser was dragged into the air by four engines, each driving a four-bladed propeller some 16 feet in diameter – to my mind a much more elegant system of propulsion, but far less powerful, and only just up to the job.

After all the seemingly insurmountable difficulties on the ground during the preceding weeks we were off at last – our course set for Prestwick, where we were to disembark for dinner in the Airport restaurant. It was still daylight, and sunny, and I left my seat and wound my way down the spiral staircase to the lounge below. My colleague was already ensconced at the bar, where there appeared to be an unlimited supply of all kinds of drinks 'on the house'. For many, this was sufficient reason for remaining below until such time as repast of a more solid kind was being offered elsewhere. For others, among whom I counted myself, it was the view from the windows that drew me again and again. Situated, as the lounge was, in the 'belly' of the aircraft's fuselage, its windows looked downwards at an angle of almost 45 degrees, making them ideal for observation purposes, and, of course, for photography – which was to be one of my main activities, whenever we were on the move.

We floated in to Prestwick on a sea of billowing cloud. Outside the windows, in the dusk, the red glow of the engine exhausts contrasted strongly with the late evening sky, from which the last vestiges of light were beginning to fade. Soon we were sinking through the cloud, and exchanging the world of beauty above them for the darkness of the night below. Straight ahead were the runway lights, and in next to no time came the dull thud as the wheels touched terra firma again, and we were taxiing to a standstill.

No apologies offered for that little excursion into the world of fancy. I am glad that I was forty years old before I had my first serious experience of flying – for that matter glad, too, that I was fifty-seven when I learned to drive a car. The excitement of both experiences is still with me in old age, but I wonder how many of the present generation of children will be able to say the same thing in their old age? From babyhood they have taken for granted so many things deserving of wonder – indeed, it could be argued that the very capacity to wonder has been largely lost. So, I stand by my image of a Stratocruiser floating on a sea of cloud, and I shall continue to enjoy the memory of being on my own in a car, driving through the heavy traffic in the centre of town, within half an hour of passing my test (to say nothing of making my first acquaintance with a computer at the tender age of seventy-five).

Dinner at the Airport was a sumptuous affair: soup, sole, chicken, fruit and ice cream, cheese and biscuits and champagne, calculated to leave us in such a state of lethargy that any anxiety about the long sea hop to Iceland was exceedingly unlikely. We were off again at 11 pm, after being told that our route would take us first to Reykjavik, and thence over the southern tip of Greenland, to Goose Bay, Quebec, and Montreal. This diversion had apparently become necessary to avoid hurricanes harassing the more southerly route. In such seemingly untoward circumstances sleep did not come easily, and it was fascinating to watch the townships float by far below, for all the world like glittering clusters of jewels.

We left the coast of Scotland behind at last, and with no more distractions, the notion of sleep could no longer be kept at bay. Across the gangway sprawled an American youth, already sound asleep, and snoring. He had boarded the plane at the last minute in plimsoles, with a rucksack on his back, looking for all the world as though he had just jumped on a bus to travel two stops along the road. I fell to wondering whether he commuted regularly across the Atlantic – it would certainly explain his blasé attitude, which I found myself envying at first. But not for long –

surely he was a shining example of those who had lost that sense of wonder, and was already taking it all for granted. The envy returned then, not for his ability to take the Atlantic crossing in his stride, so to say, but for his ability to sleep through it so soundly. For myself, sleep continued to evade me, and I was still wide awake when the lights of Iceland suddenly loomed into view, far below, at 2 o'clock in the morning. There was a nip in the air – but nothing more – as my colleague and I walked from the aircraft to the Airport buildings. We had been told that there would be a stop of three-quarters of an hour for refuelling, and we had elected to explore the possibilities of the restaurant.

We sat down at a large table, opposite a foreign gentleman who had next to no English, a linguistic status matched by the Icelandic waitress who attended us. Hot chocolate seemed to be the only drink available, which my colleague declined politely. For my part, I thought it was a most appropriate drink to have at 2 30 am in the middle of Iceland. Meanwhile, my colleague - in sign language - succeeded in convincing the foreign gentleman that he (the foreign gentleman) would also like what I was going to have. In due course, two steaming-hot mugs of chocolate arrived, and our companion opposite proceeded to inspect his with a somewhat pained expression. Somehow, then, he managed to indicate to my colleague that he considered that he (my colleague) had been guilty of an ungentlemanly deception.

My colleague, by this time, was enjoying the joke immensely, and addressing our companion in voluble English he stated that I was, in fact, a cocoa addict, and nothing would give me greater pleasure than to drink his (the foreign gentleman's) chocolate as well. In some mysterious way (and to my great misfortune), the tenor of what he had said sank in, and I was duly presented with a second mug of chocolate, which I had to imbibe as gracefully as maybe, and with as little thought as possible given to any 'bumpy' weather that might be ahead of us. I don't think I have ever quite forgiven my colleague! – but, as I look back on this little episode, I think I understand the skittish mood which had prompted it, and which had been induced by the surreal circumstances in which we had found ourselves.

It didn't stop there. Outside, in a corner of the entrance hall, the Airport shop was doing a brisk trade, a sight that, in the small hours, only added to the sense of surrealism. I bought some postcards, wrote several, and posted them. But there was more to follow. I began to scrutinize the souvenirs with which the shop was largely stocked. Soon, my eyes lighted upon a small doll, with no special attributes except the Icelandic national costume in which it was dressed. A vision of small daughter rose up before me, a thousand miles or so away already, tucked up and asleep long ago. I bought the doll there and then, and once I had it in my hands, in its rustling tissue paper wrapping, the thousand miles didn't seem to matter any more. I realized that I would have the doll with me from then on and that, symbolically, it would stand in for small daughter for the rest of my journey. What a wonderfully comforting thought to have in the small hours! My colleague was inclined to attribute my purchase to lack of experience in these matters — that I was in fact behaving like a typical rubber-necked, souvenir-hunting tourist, middle of the night though it was. He was wrong, of course, but I would have had difficulty admitting to the pain caused by the sudden, stabbing realization that, however much I might long to do so, I would not be able to tiptoe into small daughter's bedroom, and contemplate her fair head asleep on the pillow, for six whole weeks now.

'BOAC Flight 601. BOAC flight 601. Will passengers please re-board the aircraft...' From my seat in the plane I viewed the scene to which I had so briefly belonged, yet it already seemed alien, part of another world, despite the hard evidence to the contrary provided by a small doll dressed in black and white, with gold sequins. Outside, in what I liked to think of as the Arctic night, the flaps were being flapped, the elevators exercised, and the rudder rocked from side to side, prior to taxiing to the runway. The engines roared into action, quietened, and then roared into action again, and with the brakes on, the whole plane juddered, straining at the leash, as though desperate to become airborne again. Once more the engines quietened, then, suddenly, we were taxiing back to the Airport building again. The engines petered out altogether, and men were soon to be seen on the port wing. It was rumoured that petrol had been spraying out of a fuel tank - a singularly unhealthy thing to happen, it seemed, though the truth of it was never confirmed. A good deal of messing about went on, which included attacking the wing with some very large spanners. Speaking for myself, on my first journey by air, there was only one thing to do with a plane that went wrong just before take-off, and that was to take it out of service and substitute one in good order. Instead, they emptied out all the fuel they had just loaded, in order to repair a broken fuel gauge pipe – proceedings which were certainly not conducive to relaxation, let alone sleep.

It was 5 o'clock, with dawn breaking, before we eventually took off for the long hop over Greenland to Montreal, and sleep came at long last. I woke again at 7 30 am (BST), to find that the dawn was still pursuing us as we sped westwards. Behind us, it continued to unfold in slow-motion, in a riot of colour. The dark brown-grey mountains of Greenland pushed up through the billowing sea of cloud below us, the sunlight catching their snow-capped peaks, turning them to pure gold. It was a scene of timeless beauty, never to be forgotten.

Life began to stir again in the plane. A stewardess appeared with a welcome cup of tea, and some small children, whom I hadn't seen the night before, began hopping up and down the gangway as though it were a suburban sidewalk. Children take things for granted so easily, in this technological age of ours. How thankful I am that our own children, without any special guidance from us as parents, each found their own fields of interest, which daily excited them to wonder at the stupendous richness of life in all its phases. And we, their parents, were led by them to contemplate their personal and individual discoveries, which, in our time, we had passed by. It matters little whether it be the life habits of ants or antelopes, or the structure of artichokes or atoms which excites our wonder: of one thing I am certain – that to live without wonder is hardly to live at all, for wonder is the life blood of the spirit.

I washed (and showered!) in the men's dressing room, and fell to ruminating on the odd position in space I was occupying for so mundane a task: suspended in mid-air three miles up just off the southern tip of Greenland. An idle and wayward thought, but of the very stuff which makes any journey memorable, and the uncommonness of common things the more discernible.

As we crossed the wild coastland of Labrador, we breakfasted on bacon and egg, sausages, rolls and marmalade, fresh fruit and clotted cream. It was a glorious morning, with glimpses of Labrador to be seen through the billowing golden clouds far below; and above it all the brilliant emerald and blue of the sky. My camera was busy, and one 'catch' I particularly prized was that of a gleaming silver and scarlet 'Shooting Star' fighter, as it appeared of a sudden on our starboard side, and then rolled away in salute.

I joined my colleague in the lounge below, where it emerged that he had just demolished two of our fellow passengers in a voluble, not to say violent argument about the National Health Service. They were seasoned Atlantic air-travellers, and the views through the windows meant nothing to them, and they would name some particularly vague piece of coastline with a touch of the disdain which might be used by a townsman who found himself in the position of having to identify Clapham Junction for his country cousin. For myself, I am still enough of a child at heart for the experience to remain an adventure, were I to cross the Atlantic a hundred times. Familiarity breeds contempt (and boredom) only if the quintessence of the experience has escaped us altogether.

By the time we passed over Seven Islands in the St Lawrence (identified for us with calculated casualness by our two companions), the conversation had turned to Canada. Some impression of the primeval expanse of Northern Canada had come up to us through the gaps in the clouds, and one could hardly refrain from wondering what the prospects of survival would be in such terrain in the event of a forced landing, a thought which was to recur in other parts of Canada and the States. I remembered that a friend who had trained with the RAF in Canada during the war had told me that they had been given a sort of survival kit, which included such items as a fishing rod. I wondered how many fishing rods we had aboard the Stratocruiser, or, for that matter, how many of us would be able to use one to good effect if it came to it. But it didn't, of course, and whilst we were skimming the top of a cloud layer shimmering like a snowfield, which had deprived us of a glimpse of Quebec, a chatty voice over the address system told us that it was a fine morning in Montreal, and that we would be landing at 9 30 am local time. And a little later we started the long descent.

Customs and Health formalities at Montreal were simple and few. A couple of my traveller's cheques were quickly exchanged for Canadian dollar bills, to enable me to send the cable for which they would be waiting anxiously at home – 'Happy landing, Montreal.' The equivalent nowadays would presumably be a quick call on your mobile...*tempora mutantur.* A few hours later (so I was told), Roger ran round to his Grandad's as fast as his legs would carry him, and long before he reached the door he was shouting out at the top of his young voice, 'Grandad! Grandad! Daddy's got there!'

I had, indeed, arrived in a far country.

Chapter 10

It was the end of the beginning. Even though there would still be many moments of relaxation, pleasure, deep joy and even good honest fun, from now on our time and energies would be almost wholly engaged in the fact-finding that was the purpose of our six-week journey across Canada, and back across the States. To recapitulate: we were to visit a number of radiotherapy centres in North America where cobalt units had been installed, to gather information about the units, and their *modus operandi* in the treatment of cancer. And we hoped, on our return, to be able to recommend to the Hospital which particular unit would best meet our needs. Chalk River – the Canadian Atomic Energy Establishment - was also on our itinerary, Oak Ridge, the equivalent American establishment, having been removed from it in the vain hope of obtaining our visas in time for our boat passage.

A detailed account of such an undertaking would be out of place here, and, indeed, would require a book to itself! So, I must be content with a 'synoptic' account, which is bound to be somewhat kaleidoscopic - pinpointing the high spots, whilst allowing a brief mention of the handful of 'low spots' – inevitable on a journey of such length in both time and space.

From Montreal we flew to Ottawa – our jumping-off point for Chalk River, about 100 miles WNW of the city. We did find time to see a little of Ottawa, a beautiful city, and we also went over the Houses of Parliament. There were real, live 'Mounties' on guard there! - one of whom proudly posed for his photograph in his striking scarlet-jacketed uniform and wide-brimmed khaki hat. I have to confess to feeling a little guilty as I toured the Parliament buildings, having never set foot inside the Palace of Westminster (which is still true, fifty years on...)

The Atomic Energy Establishment at Chalk River was, as one might expect (for security and safety purposes), in a very isolated spot, totally surrounded by pine forests. It was little wonder that the drinking water tasted of pine oil - quite refreshing, in fact.

In the process of inspecting one of the nuclear reactors, and learning how it could be used to produce intensely radioactive cobalt sources, we were told of a horrendous accident that had occurred at the Establishment not all that long before. The basic neutron 'chain reaction' in one of the piles had run out of control as a result of human error, and in a matter of a very few seconds what has come to be known as 'melt-down' had occurred, a reference to the fact that the uranium fuel rods had become molten. Fortunately, they had only recently been renewed, so there were practically no (highly radioactive) 'fission' products present. Even so, over 2000 soldiers were employed in clearing up after the accident, each soldier allotted a task lasting only a few seconds, in order to limit his radiation exposure to acceptable limits. It could so easily have been another Chernobyl.

It was at Chalk River that we made the acquaintance of a certain Dr Cipriani. A well-known scientist, he was also quite a character, having a boa constrictor as a pet, and a Praying Mantis Vivarium in his office. There were also some pithy aphorisms on the walls: 'There are no chiefs here, only Indians' – 'It's what you learn after you know it all that counts' – 'Consider the tortoise, he makes progress only when his neck is out', and so on. We were to meet many more examples of that kind of potted wisdom during our journey through Canada and the States. We English, I suspect, are a little too phlegmatic to decorate our walls with that sort of thing – a pity.

If our eighteen-hour flight in the Stratocruiser could be said to have been uneventful, our 200-mile flight from Ottawa to Toronto was certainly to make up for it. To start with, the contrast between the two planes could hardly have been greater, the flight to Toronto being in a twin-engined, 22-seater Dakota – (even then) a 20-year-old design which had become well-known in the Second World War for its role in delivering British and American paratroopers. And there were times in our flight to Toronto when one could almost have wished one had access to a parachute. But I anticipate.

In these days it is difficult to imagine that the presence of two Englishmen among the 22 passengers was something of a novelty – but it was, and it apparently warranted our being invited to the cockpit, to meet the captain. And was he proud of his Dakota! 'She's a wonderful aircraft,' he said, 'and they'll be still flying them in fifty years' time.' On that score I believe he was right: I think there are, indeed, Dakotas flying still. What about that parachute then, that I thought I might have needed?

The fact is that, whilst we were talking to the captain, whose nonchalance certainly had to be admired, we were actually flying through a whole mass of thunderstorms, so widely distributed, he said, that there had been no point in trying to fly round them – and as the plane was incapable of flying above them he had had no option but to fly through them. I think my colleague and I were among the very few passengers who did not succumb to air-sickness – the plane having plummeted with terrifying suddenness many times, as it encountered air pockets. For me, I think, it was a case of 'ignorance is bliss'. It was, after all, only my second flight ever, and I was rather enjoying the 'firework' display through the windows, as I watched the whole length of lightning flash after lightning flash - from the base of the cloud where it began, to the point on the ground where it struck. It was a brief but eventful flight, and I must admit that though I'd escaped airsickness, I was nonetheless thankful to disembark in one piece.

It was in Toronto that we saw our first cobalt unit. The weight of the unit, with its massive lead shielding for the cobalt source, was such that it could not be installed in the existing Radiotherapy Department (which was on the first floor of the building), and an extension jutting out from the main building had been built, supported on two massive concrete pillars. The general impression created by this structure made one realize (before one had set eyes on the inside of it) that it housed something very special - and we were mightily impressed, both by the unit itself, and the kind of treatments that could be carried out by means of it.

So far as Toronto City was concerned, we had little time for sightseeing, though we did take a quick look at what we now know as a Supermarket - but unknown to us then. It was in Toronto, too, that I first heard the term 'grid-lock', and, indeed, witnessed something approaching it, on the City streets. I also remember being put down at a particular point by one of our Canadian colleagues, and being told that he would have to 'circle the block' until we reappeared. At the time it seemed an unusual, not to say extraordinary, solution to a parking problem, but it has long since become a familiar tactic in this country too. In fact, it must be the universal experience of travellers to the North American continent to become aware of all sorts of ideas and practices - both social and technological - unknown in this country, only to find, soon after, that they have been 'exported' here, and hailed as 'novel'.

Only a few pages back I said that, from Montreal on, there would be only 'moments' of relaxation. How could I have overlooked Niagara Falls? How could anyone overlook Niagara Falls?! We simply had to find time to pay them at least a brief visit, situated as they are on the south side of Lake Ontario, virtually opposite Toronto, and a mere 50 miles or so away as the crow flies. We are all familiar with the general appearance of the Falls, but perhaps a little less familiar with the awe-inspiring statistics associated with them. Their height is a modest 160 feet or so, but the volume of water flowing over them is a well-nigh incredible 1.2 million gallons per second. Such a vast flow of water falling 160 feet represents a huge amount of power, if converted to electricity. To preserve the general appearance of the Falls, the United States and Canada signed a treaty limiting the amount of water which could be diverted for power generation purposes, but even so, the limited amount permitted by the treaty has resulted in an American and a Canadian facility each producing over 2,000,000 KW. These facilities (which we inspected – and very impressive they were, too) are situated a few miles down-river, and the diverted water is ducted to them via underground conduits. Each one of a bank of turbines, I seem to remember, was producing electrical power equivalent to 100,000 horse-power.

We inspected the Falls themselves from the 'Maid of the Mist', a small, sturdy steam boat, built like a tug, which took us closer to the Falls than was really comfortable, despite the oilskins with which we were provided. But what a spectacle from close up! We also inspected them from behind - a weird and awesome spectacle – made possible by tunnelling, and the excavation of an observation platform in the rock face behind the Falls.

It goes without saying that during the whole of this trip my camera hardly stopped clicking. In this connection it is perhaps worth commenting that if one takes considerable care one can photograph the Falls in such a way that they have the appearance that one might expect them to have – a majestic and awe-inspiring natural feature miles from anywhere. But, with less care, the chances are that your photograph will appear to show a skyscraper almost skirted by the Falls – or even rearing up out of the middle of them!

Our next port of call was to be Saskatoon, halfway across Canada, and in the heart of the prairies. The plane we flew on was another Dakota, and once more I was in for a surprise. I found myself sitting next to Lady Nye, the wife of the High Commissioner. Discovering that I was English, she became quite communicative, and told me a lot of interesting and useful facts about Canada. The most memorable took the form of a little poem that encapsulated the east-to-west features of that vast country:

> 'A thousand miles of water,
> A thousand miles of plain,
> A thousand miles of forest,
> And then the sea again.'

It took just a few days to confirm that for ourselves.

Saskatoon was a bustling city that seemed to us to be in the middle of nowhere. Situated on the Saskatoon River, it had a large general hospital, incorporating a Radiotherapy

Department that had a worldwide reputation, thanks, particularly, to the work of Professor Harold Johns, head of the Physics Department. The mere mention of his name brings back a homely domestic scene – of his wife going out to collect blueberries (on the banks of the river) which later were to form the basis of a blueberry pie – highly traditional in those parts. And a very fine example of the species it was, too.

In the hospital, we inspected our second cobalt unit, and I found much to discuss with Professor Johns regarding the measurement and control of the radiation dosage that it delivered. Another quirky little memory then: we went up onto the roof of the hospital – a tall building giving magnificent views of the prairies surrounding the hospital on all sides - miles, and miles and miles of them, uninterrupted in whatever direction one looked. The vivid blue of the sky was divided up by parallel streamers of light cloud, spaced regularly overhead, but closer and closer to each other as they approached the far-distant horizon. Curious, I thought, until I realized that the regular cloud formation, combined with the totally flat and uninterrupted terrain for such a great distance, had made it possible to become aware that clouds were subject to the laws of perspective just as much as the landscape below them. Later on in my life, I found this discovery very useful when I took up landscape painting.

From Saskatoon we flew to Calgary, not because there was anything of radiological interest there, but simply because it was to be the starting point for the most fabulous train journey I have ever made - the 24-hour journey through the Rockies, and that 'thousand miles of forest', to Vancouver. The very name 'Calgary' conjures up images of cowboys and the Wild West, but a mere overnight stay on a rainy night gave us little opportunity to replace those mental images with real-life ones – always assuming that the town had retained such associations. However, I do seem to remember passing the (empty) fabric of a large cattle market on our way to our hotel. The only other memory of Calgary that comes back to me over the years is one that one certainly wouldn't associate with the Wild West days. It is of a large neon sign just down the road, and visible from our hotel, its orange-red light reflected in the rain-soaked street. It consisted of just two words – 'JESUS SAVES'...

The drizzle was still with us as we boarded the train the next morning, and it did not augur well for sightseeing from the train. The train itself was incredibly long – the number of coaches uncountable as it came to a halt in Calgary station. It must have approached a quarter of a mile in length, a supposition that tends to be confirmed by the fact that later on, in the Rockies, the front end of the train could be seen across a small ravine emerging from a tunnel that the rear end of the train (where we were ensconced) had yet to enter!

The terrain between Calgary and the Rockies was flat, but for some time any view that we might have hoped to have of the mountains was blotted out by the drizzle. Fortunately, we gradually left the drizzle behind, and there, on the horizon - and some fifty miles away still - we had our first glimpse of them. Snow-capped and, despite the distance, looking like glittering diamonds, as they caught the morning sun ahead of us. In those days the train was not one of those sleek, streamlined affairs, and at the very end of it, to our delight (and especially to mine, for photographic purposes), there was an open observation platform, surrounded by iron railings. It hardly needs be said that I spent most of the daylight hours there, taking dozens of photographs of the breathtaking scenery. Incidentally, we seemed *destined* to meet notable people on our travels, for we learnt that the elderly gentleman who was one of those sharing the observation platform with us for a great deal of the journey was none other than Sikorsky, the inventor of the helicopter.

The railway was the Canadian Pacific, whose route through the Rockies took us roughly westward through the Banff National Park to Lake Louise, and onwards through the Yoho National Park, along what must have been glacial valleys, climbing steadily to reach what was designated by a large notice as 'The Great Divide', at a height of 5352 feet. The remaining half of the journey was occupied by the long descent to

Vancouver. The final stages of this was southward along the Thompson River, until it joined the Frazer River as it swung roughly westwards again towards Vancouver.

How can one describe such scenery? I sought a single word that might characterize it, feeling as I did that despite its ever-changing aspects, there was an underlying quality common to it all. Vast? Of course - so much so that as one looked back at the track we had just passed over, it looked more like a model railway, on which one might expect a clockwork train to appear, at any moment. Primeval? Certainly - it was timeless, giving the appearance that it had always been there, and always would be, unchanged and unchangeable. A wilderness? For many, the word conjures up an image of a bare, barren and rock-strewn place, unbearably hot by day, and unbelievably cold by night – the sort of place in which Christ spent forty days and forty nights; but - for me - rather does it signify somewhere so vast and so remote that one can imagine that even in this day and age only a tiny fraction of it has borne the weight of human feet. And my over-riding thought was just that, as I looked at those endless expanses of pine forest cladding the flanks of countless, towering, snow-capped mountain peaks, or the sheer rock faces too steep to support life of any sort, other than rudimentary lichens. Apart from the puny slash inflicted on it by the single-track railway line, the vast bulk of what I was looking at, hour after hour, had yet to be set foot on, by a human being. That's my definition of a wilderness!

The descent to Vancouver was equally spectacular, following the Thompson River westwards and then southwards until it joined the Fraser River, continuing southwards then, until the river and railway took a final turn westwards towards Vancouver. The blasting and chipping from the side of the canyon to form the rail bed has been described as 'one of the great feats of 19th century railway building'. Certainly, one experienced a sense almost of disbelief at being in a train, coasting alongside a wild river torrent at the bottom of a canyon so deep and so steep. And inevitably I found myself pondering what the cost in human life had been, to make my journey possible.

The desolate nature of the terrain seemed only to be emphasized by the sudden appearance – as from nowhere - of two men on a little four-wheeled bogey, presumably petrol-driven, who caught us up in the gathering dusk. I presume they were track inspectors, or some such, and I couldn't help wondering what would have happened to them if they had run out of fuel. Come to think of it though, they could probably have free-wheeled all the way to Vancouver...

And what a sight Vancouver was to present to us! It may not have been quite so beautiful and elegant as it seemed, after all the wildness and desolation, but, in the gentle autumn sunlight and its setting against the backdrop of the Coast Mountains, it certainly looked serene – just the place to spend a day or two after the – almost literally – rough and tumble of the journey we had just made. The contrast could hardly have been greater: to be inspecting another example of one of the latest products of the atomic age hard on the heels of a journey through such timeless terrain. But for me, personally, Vancouver was to provide yet another journey back through time.

As a Quaker - or to put it more formally, as a member of The Religious Society of Friends, journeying far afield, I carried with me a Letter of Greetings from my local Friends' Meeting in Guildford – a letter which I could use to introduce myself to any other Friends' Meetings I might attend on my travels. I knew that there was a small Meeting in Vancouver, and as we were to be there over a weekend, I paid them a visit – my first such, since leaving England. And what a surprise awaited me! It transpired that the Meeting was comprised entirely of several generations of the same family, a situation that had apparently obtained for generations back into the past. I was the first visiting Friend they had set eyes on for a very long time, and if Quakers were not such staunch advocates of a classless society, I would have to say that they treated me like royalty! The whole episode had an amusing sequel. I felt moved to give some account of it to my colleague, Dr Boden, who, entirely ignorant of the ways of Quakers, questioned me in some detail about our traditions and practices, including our mode of worship. I took some time and trouble to describe how our Meetings were based on silence, which could nevertheless give rise to 'ministry', a verbal contribution on some worshipful topic which anyone present might feel moved by the Spirit to give. I was careful to point out that, even so, many a Meeting was silent for the whole of the allotted hour. My colleague indulged in a brief silence himself then, before making his response. 'Personally,' he said, 'if I feel like going to church I prefer a good show, and I go to Westminster Cathedral.' It takes all sorts to make a world, doesn't it? – and what a good thing that is!

To provide a little more sightseeing 'by the way', we took the boat to Seattle, where we were to enter the United States by the back door, so to say, and overnight we stayed in Victoria, on Vancouver Island. In spite of the fact that it is the capital city of British Columbia, it had a more 'laid back' character than Vancouver, and a certain charm which retained something English about it. It is a city well stocked with Art Galleries and Museums, and many large parks. Amongst the latter was Thunderbird Park, where there was a permanent display of Native American totem poles. It was there that I met (and photographed) Sam, an elderly American Indian, who, with a variety of woodcarving tools, was busy converting a large and very long tree trunk into a totem pole. It was highly skilful and artistic work, and he chatted to me proudly of his lifelong occupation. Another vivid memory I have of our visit to Thunderbird Park was of three little dark-haired American Indian boys who insisted on lining up to have their photograph taken – they couldn't have been more than six or seven years old, but very persistent they were, and very charming.

That day in Victoria was the far point of our journey, and from then onwards we were, in a sense, on our way home again.

Chapter 11

We had no official business in Seattle – just an overnight stay, and then a plane direct to Chicago. Certainly, Seattle would have justified a longer stay, had we had time to spare - set, as it is, amidst dramatic scenery. There are three lakes within the city limits, and mountain ranges to the west and east, the latter including the spectacular 14.000 feet high Mount Rainier. Its industries include the Boeing Aircraft Company, and long after we were there, it has, of course, become the home of the world's largest software company – Microsoft.

The flight to Chicago was, in fact, my first experience of a long west-to-east flight, and in the process we 'lost' three hours, and I well remember the quite uncanny feeling I had on arrival, when it was getting dusk in what seemed to be the middle of the afternoon. Of course, in these days of universal air travel, such experiences pass almost unnoticed – a pity, really. The city itself (the third largest in the U. S.) I found very impressive, with its 29-mile waterfront on Lake Michigan. Incidentally, it was on the waterfront, on a Sunday morning, that we came across a very large crowd, milling round something obviously of great interest. Our curiosity aroused, we went along to investigate. It was a dog show! – a very popular event in Chicago, we were told – but my colleague walked away in disgust. He was a dog-hater...

Later that Sunday morning I attended the second Quaker Meeting of our journey. It could hardly have been more different, compared with that first one in Vancouver. Held on the University campus, with dozens attending, it was followed by a very pleasant, almost sumptuous, luncheon, staged on a tree-shaded lawn adjourning the Meeting Room. I suppose - for different reasons - I felt a little out of place in both Meetings. The Vancouver Meeting was unique – so small, so 'inbred', and so isolated that one might say, in these Star Trek-riddled days, it had been 'left behind, in a time warp'. The Chicago Meeting, on the other hand, comprised a well-groomed, highly intellectual and sophisticated group. The average Quaker Meeting in this country bears little resemblance to either, typically consisting of workaday folk engaged in mundane jobs such as school teaching, social work and the like, with perhaps a smattering of medics and dentists in the larger urban Meetings – just 'ordinary' folk, in fact.

The radiotherapy centre in Chicago was indeed a luxurious one – typified by the fact that the radioactive source in the treatment head of the cobalt unit was shielded, not by what I am tempted to refer to as 'common or garden' lead, but by solid uranium! I suppose it is what one might expect in the third largest city of the richest country in the world. It did, of course, make the head that much more compact.

From Chicago we flew on, eastwards, to Cleveland. We were to land in the dark, and I had my first experience of an aborted landing. We had all but touched down, when suddenly all four engines burst into full power again. It was terrifying. One knew it ought not to be happening, yet its full significance, and why it was happening, and exactly why it was so terrifying, was not obvious. We might have guessed, of course – it was just a little matter of over-shooting the runway...

It was at Cleveland where we inspected a cobalt unit manufactured by the Picker Company, which, as it transpired, was to be the one we ultimately recommended The London Hospital to purchase. Otherwise, apart from that aborted landing, I have no special memories of Cleveland, and we were soon taking off again for Washington, where we were to visit The National Bureau of Standards, the American equivalent of The National Physical Laboratory, and the home of Dr Taylor (who had listened to my paper in Copenhagen, and had produced that wonderful aphorism about 'sittin' on the fence but not necessarily sittin' pretty').

The weather was appalling in Washington, but apparently typical – 'the two 95s' - 95 deg F, and 95% humidity. Despite that, our brief stay was both profitable and enjoyable, with useful discussions at the N B S on basic dosimetric methods and measurements (they had already built a new standard, eliminating those embarrassing errors I'd disclosed at

Copenhagen). These were followed by a lightning tour of the city's most notable buildings, including the Washington Memorial, the Lincoln Memorial, and, of course, The White House. This tour was made possible by hiring a knowledgeable taxi driver to whisk us round.

At last, then, to New York, where we stayed with friends of mine on Long Island. Again temperatures neared the 100 mark, and one thing that stands out strongly in my memory is the dreadful cold I developed as a result of going from those baking-hot streets onto the air-conditioned Subway, which felt like walking straight into a fridge in my sweat-soaked shirt.

It was in New York that we saw something of the wealthy and lucrative private sector of U. S. medicine, where a course of radiotherapy in the private sector in 1954 cost the equivalent of something in the order of £10,000 in these days. It reminds me of what one reputable Canadian radiotherapist said to us, speaking of the state of the art in the U.S. 'They don't need physics or physicists in the U.S. - you see, they treat by ear.' We should not, of course, take such a remark too seriously! Another quick tour round the City sights took us, among other places, to the top of the Empire State Building, from which vantage point we unexpectedly caught sight of the 'Queen Mary' in harbour; she was to carry us in leisurely and luxurious style on the last leg of our journey. Inevitably travel-weary and very homesick after six weeks away, at the very sight of her I lost any remaining appetite for foreign travel that I might still have had: my one desire was to get on board her, and wave 'Goodbye' to the Manhattan skyline as we watched it fading away. The novelty of foreign travel had definitely worn off.

We had become accustomed to the confined space within an aircraft – and confined it certainly was, even in the Stratocruiser - despite its upper deck, and the lounge below - and one's response on boarding the 1000ft long 'Queen Mary' could be nothing other than one of sheer disbelief. Again I need to emphasize the novelty of the experience for me: as I stepped on board, it was to recollect that the largest ship I had previously travelled on was a ferry to the Channel Islands...

As we sat at table in the lavishly decorated dining hall, it was difficult indeed to believe that this vast space was hurtling along at something like 35 miles an hour as the ship – all 100,000 tons of her - ploughed her way through the Atlantic Ocean. I remember hanging over the stern rail one day, watching the churning fury of the ship's wake as it left the propellers. I had gathered from the official literature that the ship's turbines developed a total of 200,000 horse-power. I had a somewhat intriguing thought then. At Niagara, it took just two of the turbines I saw there to convert the kinetic energy of water from the falls into electrical power equivalent to 200,000 horse-power; and what I was witnessing as I leaned over the stern rail was the exact opposite – 200,000 horse-power turbines producing the equivalent amount of kinetic energy in the water being propelled rearwards from the ship. An interesting comparison? Well, anyway, it was for me!

As we boarded the 'Queen Mary', little did we know that our growing habit of bumping into well-known people on our travels was to produce its most notable example of all. Out of curiosity, I went through the passenger list, and to my great excitement I found that it included one Martin Niemõller. It is a name not nearly as familiar to the present generation as to mine. He was a pastor in the German Lutheran Church, who, after a brief flirtation with the Nazis in the early 1920's, broke with them very early, and became a leading figure in the Church, among those fiercely opposing them. He was eventually arrested for treason, found guilty but, probably owing to world opinion, given a suspended sentence. He was immediately rearrested on Hitler's personal orders, and, until the end of World War II, was held first in Sachsenhausen, and then in Dachau. Towards the end of the war he narrowly escaped execution. What a man for a Quaker to have the opportunity of meeting in person!

I dropped him a note immediately, via the ship's post, and received a quick reply, suggesting we had tea together. My colleague, normally a fairly imperturbable man, was visibly moved by the occasion, and I did have a photograph of the three of us at tea, taken by the ship's photographer, but, to my chagrin, I have been unable to locate it.

That tea party was one of the most memorable occasions in my life: it was certainly, for me, the most memorable event of our six weeks' journey. And I think that went for my colleague as well as myself. Our guest had been to a World Congress in Canada, which we discussed at some length, though I can remember little now of the substance of the discussion. Indeed, the Pastor was as keenly interested in what we had been up to, as we had been to get him talking about himself. That early flirtation of his with the Nazis no doubt owed much to the fact that in the First World War he had actually been a U-boat commander, and it was evident from our growing acquaintanceship with him during the rest of the voyage that the sea was in his blood. He himself confirmed this as we watched the ship being docked in Le Havre harbour. He turned to us and said, with a glint in his eye, 'You know, give me 24 hours and I could navigate this boat with the rest of them!'

He lived on, continuing to play an important role in the international affairs of the Church, and died in 1984 at the age of 92.

What a journey we had had! – for me, truly the journey of a lifetime, and that is why I have done my best to do it justice, in the preceding pages. And what a homecoming! Another photograph which as yet I have failed to locate was taken with the self-timer on the camera – of Mary and me in our back garden in Guildford, within minutes of my arriving home. Although the photograph itself has proved elusive, the image recorded on it certainly hasn't – as fresh in my mind's eye as though it had been taken only yesterday. It was of two people wearing such broad smiles which – yes! – really did spread almost from ear to ear.

Chapter 12

It is a well-worn saying that moving house is second only to bereavement on the trauma scale, and our move from South Ruislip to Guildford in 1949 certainly bore out the assertion. Logistically, moving house is the biggest operation any of us are likely to be involved in, in ordinary life, and there is plenty of scope for something to go wrong, and the move to Guildford was no exception.

We were almost literally about to climb into a taxi with the three children, to follow the furniture van to Guildford, when we received a phone call from the house agents, informing us that our solicitor had failed to complete certain bits of the paperwork, and that in consequence we had no legal rights to take over the property. I was totally fazed by this, and immediately rang the solicitor, who assured me that if there was any hiatus it was at the Guildford end, and advised me to proceed, and to assert our rights on arrival. That was all very well, I said, if I were the only one involved, but I was not at all happy doing so, accompanied by my wife and three children and a van-load of furniture. What if they were adamant? They have no grounds for being adamant, he assured me – *go!* So off we went.

I can still see that furniture van just ahead of us, all the way to Guildford. It was Saturday, and it was around midday when we arrived outside the house agents' office. Well do I remember how vulnerable I felt, with my whole family and all our worldly goods parked on the highroad in the middle of Guildford, possibly to be told that we had - both literally and metaphorically - reached the end of the road, and there just wasn't anywhere else to go. I remonstrated vehemently with the agent as he expressed his reluctance to hand over the keys. I repeated what I had been told by our solicitor, but he was unyielding As of that moment, he said, we had no legal right to the property. I asked what he thought I should do with my family and the vanload of furniture. Gradually, then, he relented. But I first had to sign a piece of paper agreeing that I had no legal right to the property, and that permission for me to enter it was 'on sufferance'.

So it was that on that April Saturday in 1949 we took possession of our new home, a dear little semi-detached red brick, country-cottage-style house, built in the late 1920's or early 30's. How thrilled we were with it, even though we dare hardly allow ourselves to feel that it was really ours. Believe it or not, it was the first time Mary had set eyes on it! You see, houses for sale were still as scarce as gold dust in 1949, and thereby, as they say, hangs a tale.

The Radiotherapy Department of The London Hospital had some facilities at one of the Guildford hospitals – St Luke's – which involved my going to Guildford fairly regularly, though not very often. It was in that way that I came to know, and very much to like the town and its environs, and it was during one of my official visits that I learnt that 17 Thorn Bank was on the market. Before returning home that day I got to look over the house, just ahead of two other prospective purchasers, and when I told the house agent that I liked it very much, he promptly said, 'You'd better buy it then, because if you don't one of these other two will.' I rang Mary, explaining the position to her, and, typically, she said, 'Go ahead, then!'

But as one might expect, there was more to the agent's hurry to dispose of the property than met the eye. Within days of our taking possession in fact, he was threatening us with eviction, his behaviour anything but professional. Petrified at the prospect, I rang our solicitor immediately - at his home (it was evening, and outside office hours – but I felt justified in doing so, in view of the threat we were facing). I was told by his wife that he was not available: 'He's at a Freemasons' dinner,' she said. I'm afraid I am one of those people for whom the mention of Freemasonry is a bit like a red rag to a bull. I explained my dilemma, and though she remonstrated, she finally gave me the telephone number of the hotel at which the dinner was being held. I rang the number, and explained to the manager that I had an urgent need to speak to one of the guests. It was his turn to be petrified – in his case at the very thought of breaking in on a Freemason's dinner in order to extract one of the guests. 'The very idea of it,' you could almost hear him muttering. But I insisted, and in a minute or so I found myself exposed to the full fury of our solicitor. 'I was at a Freemasons' *dinner*,' he raged, 'and you insist on getting me out of it for some insignificant little business matter...' My reply was incisive. 'Yes, I've got you out of your dinner for a matter of a minute or two, but the house agent here says he's going to get me out of my *house,* lock, stock and barrel, for good.' 'I've already told you,' he fumed, 'whatever the trouble is, it's at his end, not mine.' 'He insists that is not so, and I assure you, he means business. Will you please revue the matter, as one of great urgency.' Angrily, and very reluctantly, he agreed to do so, 'one last time,' he said, and slammed the phone down. Late the next day he rang me, as contrite as he was capable of being: the fault had lain with his clerk, after all. We breathed an almighty sigh of relief, and, for the first time, allowed ourselves to feel that 17 Thorn Bank really was going to be our new home for the foreseeable future.

On Sunday, the day after our arrival, I went for a brief walk up onto The Hog's Back, on the northern slopes of which our little garden village had been built. In little more than five minutes I was surveying the almost breathtaking vista to the south – mile after mile of it. I hoofed it back home again and burst through the back door. 'Mary! You've no idea where we've moved to!', I exclaimed. I explained to her what had happened, and before the day was out I was minding little daughter, whilst Mary and the boys went to take a look for themselves.

Talking of 'before the day was out', I have to record an amusing aspect of that Sunday – a very puzzling one, too, at the time. With no actual intention of doing so, but as gauged by the activity of our neighbours and people going by in the road, we had apparently got up late - or what seemed to be a more likely explanation, the daily routine of the 'locals' was out of step with ours. And so it surely was! Remember – we had moved only the day before, and hadn't bothered to fix up a radio immediately, and we had no Sunday paper. Yes – we*'d forgotten to put the clocks forward.*

We suffered another embarrassment in those early days. Beyond the hedge at the end of our garden, and approached by a little wicket gate, was a piece of 'waste' ground, which had been divided up into little allotments, one of which abutted neatly onto the back of our property. There were some nice fresh vegetables growing on it, too − cultivated, I naturally assumed, by the previous owner of our house. I began to help myself to some of the produce - but what I didn't know was that it was being cultivated by our next-door neighbour... Need I say any more?

And what about the house, and that house agent? We haven't finished with them yet. The house, in its general design and layout, was delightful, and all that we could have hoped for as a home. But, what about its fabric? Therein lay the rub. Superficially, we thought how lucky we were, as the previous owners seemed to have newly painted the whole of the house, inside and out. Strange − and generous, we thought. That is, until one day, quite early on, my finger went through the bottom rail of one of the bedroom windows, when I was trying to open it. There was wet rot in it, which had been concealed by that 'generous' lick of paint.

I can't remember now how I uncovered the racket that had been going on, involving a speculative financier working in collaboration with the house agent. Nothing very complicated: the house agent would become aware of a would-be seller whose property, on inspection, was found to have been badly neglected − maybe because the owner was too elderly to have kept it in good order himself, and too short of cash to have been able to pay to have it done. So , with the owner desperate to sell, the financier would offer to take it off his hands at a knockdown price. Smartening it all up then with that lick of paint (and incidentally concealing any serious defects), the property was then put on the market, and a would-be purchaser hustled into a deal, with no time to get the property surveyed. But couldn't one have insisted? Of course, but only to be told that there were other purchasers ready and waiting who were quite prepared to forgo that luxury. But why the continued haste after ensnaring a purchaser? That was the subtle bit. There was a law (I've no idea what it was called, or whether it still exists) which exempted the seller from the capital gains tax, provided the sale was within a certain maximum time following the purchase date (was it three months? – or maybe even less). Otherwise the tax would make a nasty hole in the profits. Ingenious - the victims just fell into the house agent's lap, and the financier did the rest. A pretty unsatisfactory racket to have got caught up in, one might say. Actually, I got a lot of satisfaction out of it, as things turned out: the hiatus caused by my solicitor's clerk delayed the sale just long enough – yes! – for them to have to pay the tax...You see what I mean about pre- and post-removal stress syndrome?

And I've just remembered another bit of peremptory behaviour on the part of the house agent to which, had times been normal, I would have taken great exception on the spot. His immediate response to my decision to purchase was to ask me how much I earned – just like that! Again, if I had demurred, saying that it was none of his business, he would no doubt have shown me the door – all part of that master plan of making sure there would be no hidden obstacles to a quick sale. As it was, I well remember how smug I felt when I said, '£800 a year.' – that would be between £40,000 and £50,000 in these days, I suspect. Remember – by then I had become Head of Department. It was certainly a good salary in those days, but by no means excessive: hospital work – particularly if you were 'lay' (i.e. non-medical) - was by no means generously rewarded. That is probably still the case, even today.

At last, then, we were able to settle ourselves down in our cosy little semi-detached country cottage, with no more swords of Damocles hanging over our heads. It hardly needs saying that first and foremost among matters demanding immediate attention was a school for John and Roger. It also hardly needs saying that the one thing I had made sure of on that fateful day of decision to purchase the house was the availability of a good school. What I learnt was that within five minutes' walk of the house there was a small, private Junior School with an excellent reputation, and an early interview with a certain Mrs Jarman, who was the owner and headmistress, convinced us that we would be happy for John and Roger to become her latest pupils at the beginning of the summer term. Rosemary would follow them, five years later.

With only three teachers to cover 6 years of schooling, it would seem that the arrangement of the classes would need to be unusual, with the quality of the teaching consequently suffering. The arrangement of the classes was certainly unusual, but the examination results, and in particular the number of scholarships gained at the 11-plus examination, was evidence enough that despite the unusual arrangement of the classes the quality of the teaching was of a very high standard. What about that unusual arrangement of the classes, and the way they were taught? Obviously, it involved pupils in a lot of private study at the back of a class under instruction, which would seem to be very disadvantageous - and so it would have been unless it was all very well organized and under tight control. But, 'The proof of the pudding...' - and we were more than happy that all three of our children got those coveted scholarships to a Grammar School. It wasn't as a result of concentrating on the three R's either. Discussing with Rosemary what recollections she had of the school, she told me that every Friday they were read a story from Greek mythology, adding that it was that which had given rise to her interest in the Classics, and had resulted in her choosing to do Latin and Greek at the Grammar School. I am sure that those Friday morning stories had no particular missionary zeal behind them, with the intention of swelling the numbers of those who would go on to study the classics, but Rosemary's response to them illustrates how significant something can turn out to be that had appeared to be quite incidental.

One of John's recollections could be viewed in that light, too, though it was not directly related to Mrs Jarman's school. He said that it was when he was about 9 or 10 that he had read a book entitled 'Animals without Backbones', and he added that it had changed the course of his life. Asked what he meant by that, he said that in the broadest possible terms it had got him interested in the science of life, and that for most of his Grammar School days he was looking to become a zoologist. Organic chemistry won the day eventually, and his working life was spent in drug research.

As for Roger, his only - but very strong - recollection of his primary school days was of country dancing on Monday afternoons – 'bouncing up and down on wooden floorboards,' as he put it – which, he said, invariably resulted in a headache every Monday evening! As far as early influences on his choice of career were concerned – well, there weren't any, and he chose architecture as late as his last year at Grammar School. Like me, he was a pretty even mix of science and the arts, and was torn between the two. There was a summer's evening, however, when we were out for one of our usual late evening walks, and discussing his dilemma as we went. I well remember turning to him suddenly and saying, 'What about architecture then?' 'Cor! *Yes!*' he said with hardly a hesitation. It was a dramatic moment, the occasion so deeply etched on my memory that I can even recollect exactly where we were when that little exchange took place: it was halfway along the footpath that took us from our little cul-de-sac up to the main road – but, that evening, for Roger, it went much further...

Thorn Bank was in Onslow Village, a small garden suburb of Guildford, and in the heart of it was All Saints', an Anglican Church built in an attractive, modern architectural style – in particular the roof shape was that of a 'hyperbolic paraboloid'. In a nutshell, this meant that the shape of the roof gave it an intrinsic strength such that it needed no supporting columns at all inside the building – a valuable feature in a church, since it meant unobstructed views throughout, and from anywhere in the building. I have just asked Roger why I haven't come across any more examples of what would appear to be an important architectural technique. His laconic reply was that it was something of a novelty around the 1950s or so, adding that 'novelties come and go'...

Despite our leanings towards Quakerism, we were drawn to this little church, situated as it was, in the middle of a small local conurbation in which the community spirit was very strong. We had always believed that where possible one should worship locally, in fellowship with one's neighbours, and so it wasn't long before we had got in touch with the vicar, The Rev. Mr Orr - a devout Anglican, if ever there were one. The upshot of this was that we were soon attending All Saints' regularly, making close friends with many of the congregation, and by 1952, I think it was, I was attending Confirmation classes (Mary had been confirmed in childhood). Those confirmation classes became, in fact, a turning point in my life, but not in the way you might suppose.

The classes had gone on all winter, and the Confirmation Service was fast approaching, when a problem began to loom for me, which I soon discovered couldn't be swept under the carpet. What was it that happened when the bishop laid his hands on me (I was asking myself) that changed me from someone *not* acceptable at the communion table into someone fit and able to take communion? I asked Mr Orr, earnest and deeply sincere Churchman that he was, steeped in the Thirty Nine Articles and so forth. And it was his answer that perturbed me. 'It's not that you're any different – it's – it's just a rule of the Church,' (or words to that effect). 'But if it doesn't change anything in me, it must mean that I am just as fit to take communion before being confirmed, as after.' 'That's true. But, as I say, it's a rule of the Church.' I explained then that confirmation could have little or no meaning for me spiritually, and that I could not go through with it on the basis of it being a mere legalistic formality. Poor Mr Orr! He was very upset, and he actually had an audience with the bishop on my behalf, and subsequently came to see me, looking very pleased with himself. 'It's alright,' he said. 'The bishop says you can take communion before being confirmed, provided that you give an undertaking that you will be confirmed.' I was mortified. I could see that he genuinely believed he had solved my problem, but he went away sorrowing when I told him that I was afraid that to me that was just side-stepping the fundamental issue I had raised, and that, whilst thanking him for what he had done for me, I still didn't feel I could go through with the confirmation. And there it was left.

So how did that change my life? Well – I subsequently got in touch with the local Quaker Meeting, which we soon started attending regularly, and which, in due course, accepted us into membership. It was a lovely old Meeting House – something like a couple of hundred years old, as far as I can recollect - situated in a quiet side street in the centre of town. Among its members there were some 'weighty Friends', too, to use a time-honoured Quaker phrase. One of these was Reginald Smith, a journalist who certainly had a way with words, and spoke in Meeting ('ministered') frequently, with great effect. Another was one Robert Pearl, a horticultural expert by profession, steeped in Quakerism, and I have one amusing memory of him. He was not averse to attending Evensong at the local Anglican church (he lived in a neighbouring village), but would sometimes find that the words of a hymn (like 'Onward Christian Soldiers', for example) clashed with his Quakerism. He would hastily consult the Metrical Index at the back of the hymnbook then, to find words that he could sing to the same tune without feeling that he was betraying his Quaker principles. I must say that, on occasion, I have taken a leaf out of Robert's book...

For those not familiar with Quakerism a brief account of its main tenets might not come amiss, and will, in any case, help in understanding some of my experiences as a Quaker. First, then, why that rather strange name? It's far less of a mouthful, of course, to say 'I am a Quaker,' compared with 'I am a member of The Religious Society of Friends'! But why 'Quaker'? Some would say that it dates back to George Fox's injunction to his followers 'to tremble at the name of the Lord', whilst others would say that it sprang from fits experienced by worshippers when moved by the Spirit. It is of no great consequence which of these hypotheses is correct - it's probably a mixture of the two – but whilst I have not experienced a 'fit' as such when ministering (praise be!), I have certainly experienced 'quaking', and to a quite serious, and, indeed, untoward extent, from time to time. My own experience in this matter I regard as being psychological, rather than truly spiritual in origin. Nervous, and far too self-conscious at finding myself on my feet in the midst of such an august gathering, I used to tremble – yes, literally quake – to the point, sometimes, when I thought I was going to faint. It became so serious that I chose to sit near the door – providing me with a quick escape route if it all got too much for me. Eventually it did indeed, and in 1965, with a new, young vicar installed at All Saints', I returned to what I called 'the anonymity of the pew'. It was to be nearly twenty years before I felt able to return to the Quaker fold again. But I digress.

As a pacifist, I was very much at home in The Society of Friends, and it so happened that we joined their ranks at a time when a crucial development was taking place in the weaponry of war – the development of the so-called 'hydrogen' bomb, based on the energy released in nuclear 'fusion' rather than 'fission', and orders of magnitude more powerful than the bombs dropped on Hiroshima and Nagasaki. As a Fellow of The Institute of Physics I was asked to review for them a book called 'Our Nuclear Future', co-authored by one Edward Teller, nicknamed 'The father of the H Bomb', and a German co-opted by the Americans after the war to help develop their nuclear weaponry still further. I was so incensed by this book that I was moved to write a poem about its subject matter, poetry, for me, being a far more powerful use of words than scientific prose. The poem was published in the Quaker weekly, 'The Friend', in the early 1950s, and is my earliest surviving poem. Here it is:

God's Atoms

Neutron-sundered nuclei,
chain-reacting
with their neighbours,
vindicate Einstein,
and convert
a handful of matter
into a city-scything
flesh-searing,
man-melting
fireball,
paling the midday sun.
Suspicion-split,
fear-fused,
they rear their mushroom heads
to high heaven: fungi
growing on a hot-bed of hatred
in the cellars of men's minds.

A-bombs or H-bombs,
they are yet God's atoms,
fission or fusion,
still his power, perverted
by Man's continued failure
to take his brother by the hand.
Thus God's great blessing,
atom-locked,
and burgled out of time,
must, under God,
yield true increase,
or, man-mismanaged,
gene-mutate his future
to a nothingness
of lunacy,
gaped out in deserts
sterilized by atom's rays
that might have served
God's ends.

Strong stuff – a young man's poem, you might say, but I feel just as strongly about it today. When I elected to take up physics as a career, one reason was that I saw its potentialities to benefit Mankind; but to find that there were those who were prepared to debase those potentialities to produce the most horrendous weapon of war ever conceived by man, made me as angry – perhaps more angry – than I had ever felt before. To say my revue of the book was 'scathing' is the understatement of the century.

I have to admit that, to date, my forays into poetry have been somewhat spasmodic, usually touched off by something that has stirred me very deeply. The first poem I can remember writing was when I was about 15 – for the School Magazine - and it got me an interview with my English master, who was encouraging, and said I should write more. In view of that, I wonder what he would have thought of my average of only a little over one a year to date? In fact, 'God's Atoms' is the earliest one to survive to the present day, and that was written some twenty years after leaving school! However, recollecting it at this point in my story has made me realize that the poems are part of that story, to be included where they are relevant, and of help in its telling. You have been warned!

Nevertheless, the creative urge – which has driven me throughout my life – has, in fact, by no means been confined to science and poetry. As noted earlier, before my age had reached double figures I was drawing groups of parallel lines across a piece of paper and decorating them with black blobs, and, asked what I was doing, I would say, solemnly, that I was 'writing music' (the very appearance of music committed to paper fascinating me almost as much as the sound it gave rise to!). But, during that handful of piano lessons I had around the age of 10, I learnt to 'read music', and, from then on, frequently strove to compose it. Such was my enthusiasm in this matter, that my mother gave me a book for my 16th birthday called 'Form in Music', by one Stewart Macpherson. An erudite book, describing the structure of music, from a simple minuet to a sonata or a symphony - and I still have it on my shelves, in good condition, after all these years. And how I have treasured it!

It must have been close to that birthday that - at no-one's behest - I wrote a precocious 15-foolscap-page essay on 'Music, and the Mood in which it is Composed' dealing, among other things, with the mysterious fact that the happiest of music was often written when its composer was at his saddest. Yes – I was 'writing' (to the best of my ability!) quite a bit of music by then, and evidently with some success. Do I mean by that that I got 'published'? Well – yes, and no. 'Stop prevaricating,' do I hear you say? Really! – I'm not.

It must have been in the early nineteen thirties - the heyday of popular songs like 'Goodnight, Sweetheart'. Hoping to make my fortune, I wrote a song – words and music – in the standard 'pop' format of the day, and submitted it to several publishers in Denmark Street – London's 'Tin Pan Alley', but apparently without success. I say 'apparently' for good reasons, because, one day, maybe a whole year later, to my near disbelief (when I was listening rather idly to the radio) I suddenly heard it come over the air. Yes - my song! Well - not the whole of it, because the words had been changed, but the music was mine, *in toto*. Unfortunately, as I have said, I had sent the song to several publishers in turn, but had kept no records of the firms involved, records which, to carry any weight, would have had to have been formally witnessed. I could do nothing about it, and it is easy to imagine the chagrin, and sense of disillusion that I experienced, as a mere lad in my late teens. It seemed to me that it was one thing to steal someone's cash, but quite another to steal the fruits of his creativity. That seemed to be a lot meaner to me, and still does... Nevertheless, having mentioned cash, it's worth remembering that if my song had been accepted, and published, it would probably have brought me in hundreds of pounds - the equivalent of thousands these days, and that could have made all the difference to my parents, who had made real sacrifices to enable me to have a university education.

So why didn't I try again? I don't remember having done so, and the reason was probably two-fold: I lost heart, when confronted with such blatant dishonesty in the music publishing industry, and I lost interest in the world of 'pop' music, as my involvement with classical music deepened. I must, indeed, have reached the point when a success in this area would have become an embarrassment which I would have needed to live down!

Intriguingly enough, I can still remember the first four bars of the melody line, together with the words that had been fitted to it, but I've no recollection whatever of the original 'lyric'.

Actually, believe it or otherwise, it was not my first experience of being 'pirated'. A year or two before, that insatiable drive of mine to create had had me trying my hand at short story writing. It was called 'A Fortune in Brass', and the central character was an absent-minded bank manager who tucks a valuable Bill of Exchange into one of a pair of brass vases in his study, as a 'safe' temporary repository, and proceeds to forget that he has done so. Meanwhile, his wife gives the vases to the annual village Jumble Sale. Unable to confess the loss to his wife, he behaves in an increasingly tormented fashion, much to her consternation. 'To cut a long story short' as they say, all is solved when the purchaser of the vases turns up with the missing slip of paper.

Many months after several attempts to get the story published, I was idly browsing my way through the magazines in the waiting room of my dentist's surgery when, with a sense of utter disbelief, I found myself reading, word for word, the selfsame, awkward, and guarded exchanges between wife and husband. Of course, the title had been changed (to 'Two Brass Vases') though the vases, sold to a second-hand dealer, were still at the centre of the story – and in the innocence of my late teens I had failed to spot that that little piece of paper would, so to say, have much more mileage in it as a compromising love letter.

Once again, in my teenage naiveté, I had failed to keep a proper record of the publishers to whom I had submitted the story. In fact, the magazine in which it had appeared was a very reputable one. So what were its proprietors up to, pirating material? I could only assume that they weren't, and that my work had been pirated by a dishonest reader for another magazine.

My whole life has been peppered with similar creative efforts and activities 'on the side'. For example, the early weeks and months of my Honours degree course in Physics found me busy (in what time I could spare) with the early stages of writing the music for a musical - a fellow student, one Arthur Gill, 'feeding me' with the libretto, as and when he had the time to add to it. Just one song ('Goodbye – just for a While') has survived, and, renewing my acquaintance with it after many years, I have to say that I am not at all embarrassed... I have only the vaguest of recollections of the story line: that of village life, and, in particular, the strange goings-on of the rural district council! Alas – I was no Sullivan. And Arthur Gill was no Gilbert: years after our time at King's College I discovered that he had become a well-known professional photographer.

As the years went by, what else did I try my hand at, apart from that thin thread of poetry, woven into the texture of my life? Well, it seems to have been writing plays for radio (and, some years later, for television). This was in my late thirties - a few years after the war - when radio and television were happy hunting grounds for would-be playwrights. As far as the radio was concerned, my main effort went into a full-length play, the title only of which has survived -'The Virus', inspired by the arrival on the medical scene of the electron microscope. This made use of the wave properties of electrons, and enabled much smaller objects to be visualized than was possible, using the conventional microscope - the effective magnification of the electron microscope being some 1000 times greater. This made it possible for viruses to be 'seen' for the first time, and the play was about a young scientist who managed to isolate and visualize the virus responsible for a devastating epidemic.

It would be some fifteen years later before I tried my hand again as a would-be playwright – for television this time. The title of that play was an intriguing one – 'The Essential Anguish' – and I still have a copy of it. However, the story behind the title, and my motives for writing the play, belong to the later 1960s.

Chapter 13

Despite the backdrop of such epoch-making events as the development of the hydrogen bomb, family life went on more or less normally. In 1953 John won a scholarship to the Royal (or King Edward the Sixth) Grammar School in Guildford, and I believe that it was in that same year that my parents finally uprooted themselves from Stevenage, to come to live in a bungalow in Friars Gate, a road that was back to back with Thorn Bank. The move certainly gave them new life in more senses than one, not least by being in touch with our growing family. Roger, in particular, frequented Grandad's workshop, fascinated by the multiplicity and range of tools comprising the stock in trade of an expert cabinet maker. In fact, when Grandad died, it was Roger who inherited them. He spent many a happy hour with his Grandad in that workshop of his, as he continued his trade as a hobby – and a very practical one it was, too - from time to time designing and constructing very useful and elegant articles, from fitted shelves to 'customized' pieces of furniture.

Then there was the day when Roger came running back home as fast as his legs would carry him, clutching something to his chest as he did so. 'Look what Grandad's given me,' he exclaimed, as he burst through the door. 'What is it?' we chorused, infected by his enthusiasm. 'A violin!' He was almost screaming with delight by now.

Actually, his enthusiasm was a bit of a mystery. Until then, it had been John who had shown some interest in music – the guitar, actually – and together we had constructed his first guitar from a 'Hobby's' kit. But there, Roger already had an eye for beautiful shapes, and I suspect it was the sheer beauty of the instrument's appearance that had captivated him. Nevertheless, we found a teacher for him – a Miss Dunn – and over the next few years he progressed steadily through the Associated Board examinations, emerging eventually with a credit in Grade 8. In the meantime he won a scholarship to Farnham Grammar School, where he became leader of the

School Orchestra, and I am not likely ever to forget the last School Concert in which he participated – playing Vaughan William's 'The Lark Ascending', with the orchestra accompanying him. One would have thought that this was for him the beginning of a lifelong love affair with the violin, but, early on in his University days, the affair ended, his new love the clarinet - and jazz. Later on, that affair ended too, and he became a passive listener, to my great disappointment. It isn't as though his skills as a violinist were mediocre: they were not. Somehow, classical music had failed to capture him. In fact, it was Rosemary who became the musician of the family – but more of that a little later.

Meanwhile, John had taken up his scholarship at The Royal Grammar School, and was thoroughly enjoying the academic environment. I say that with some care and a degree of preciseness, since sport – an important part in the curriculum of such a school – was, and has remained ever since, anathema to him: he just couldn't see the point of it, and throughout his life has regarded it as an unimaginable waste of time. One need say no more about his academic leanings than to record that he took five A-levels – Chemistry, Physics, Zoology, Botany and Mathematics. He does admit that he had, as he put it, 'two goes at them,' to achieve good passes in all five.

As previously noted, his passion was the study of living creatures – the science of Life itself, one might say. Two recollections of mine will suffice to illustrate the enthusiasm with which he pursued his passion. The family was out for a walk one summer's afternoon with two close friends of ours (Alec and Constance Buckels – more of Alec in a moment). John, for ever peeping and prying into the long grass and the hedgerows for whatever might be lurking there, spotted a grass snake. These are very easily disturbed, and consequently very difficult to catch. I shall never forget the sight of a twelve-year-old boy poised, like a statue, a few feet from the snake, and moving almost imperceptibly towards it. He must have taken a minute or two, before he finally pounced. The 'sting in the tail' of this story - if I may put it that way! - is that the main defence of an endangered grass snake is to eject a foul-smelling jelly – which it did, all over John's hands. So thrilled was he with his prize that I really don't believe he even noticed the smell. But the rest of us did! - and we made him walk about fifty yards behind us for the remainder of the outing. Incidentally, Alec Buckels was a very well-known wood engraver, examples of whose work are in the Victoria and Albert Museum, and who had been a pupil of the eminent Thomas Bewick. For quite some time John became a pupil of Alec's, but, unfortunately, his interest faded: the fastidious and meticulous nature of the medium fascinated him, but he had little or no interest in it artistically.

My other recollection belongs to his time in the Sixth Form. For homework he had been given the task of dissecting a rabbit, which he was performing on a board lodged across the bath. After some time had passed, he came downstairs, speaking with obvious excitement. 'Come and look at this!' he exclaimed, beckoning us towards the stairs. Starting from the opened-up rabbit in the bathroom and trailing round the landing, and at least halfway down the stairs, were the unravelled intestines of the creature. He could hardly have made the point more graphically.

As for Roger, he was nothing like as single-minded as John in his interests, and, as already observed, he had one foot firmly planted in the scientific camp, and the other in the arts. Very early on in his time at Farnham Grammar School he began to show great artistic talent, and was fortunate to have a very gifted art teacher – a certain Mr Wills, who soon realized that Roger's was no ordinary talent. Even at that age, he showed great interest in painting from his imagination, rather than from a scene or a sitter, despite the fact that he was already showing great skill as a portraitist. Hanging in my hall to this day, I have a watercolour painted from his imagination when he was in the 6th Form. It is of a street of tall buildings on a rainy night, some of the windows lit up, the light from them reflected in the wet street below. It is an extraordinarily atmospheric painting, which, over a period of some 50 years, has never lost its charm or fascination. Before he left school he had begun to find his own particular territory to explore in his painting: abstract colour compositions, and still-life paintings in which he ignored the original colours of the objects making up the still-life, and used their shapes as the framework of a colour composition.

Rosemary's speciality depended on a single fortuitous event for it to become as clear-cut and final as John's. She tells me that somewhere about the age of eight she happened to hear a flute recital on the radio, and from then on her aim in life was to play the flute professionally. Sadly, in those days, there were no flutes specially designed for children to play, and consequently she had to bide her time as patiently as maybe, until she was big enough to play a full-sized instrument. In the meantime she quelled her impatience as far as possible by playing the recorder sideways on! Her long wait came to an end three years later, when, having duly passed her Eleven Plus, her mother asked her what she would like as a celebratory present, and, without a moment's hesitation, she replied, 'A flute, please.'

Guildford, in fact, boasted a School of Music, where a certain Mr Dixon taught a whole gamut of wind instruments, from flute to saxophone. Such was Rosemary's progress with him that after two years he felt the need to hand her on to a professional flautist. Fortunately there was one such within easy travelling distance of Guildford – at Surbiton, about fifteen minutes' journey by train. He was Christopher Taylor, principal flautist in the Covent Garden Orchestra, and well known nationally as a soloist, both on the Radio, and in London, and elsewhere. After two years with him, he obtained an audition for Rosemary with Geoffrey Gilbert (who had been his own teacher), and she was duly accepted, which meant regular trips to London for two years, her time with him brought to a premature end when he departed for America.

During the 1950s and the early 1960s, when the children's education was developing and broadening out as just described, family holidays were regularly being taken. There was, of course, the notable one in Plymouth, in 1954, preceded, and followed, by others, even more notable. One, in particular, stands out above all the others, made possible by a particular event in my professional life.

Following the fact-gathering journey across Canada and The States, The London Hospital purchased the first high-power radioactive Cobalt unit to be installed in this country (followed shortly afterwards by St Thomas's). Consequently, my physicist colleague at St Thomas's and I came to be regarded as somewhat expert in that particular field, and in 1961 we were invited by The International Atomic Energy Agency in Vienna to spend a fortnight at the Agency drawing up a document outlining the principles involved in the installation and use of such potentially dangerous equipment. It was a wonderful experience, and, apart from all the hard work involved, we were able to attend orchestral concerts, and an opera, at the famous Musikverein. More importantly, perhaps, from our families' point of view, we took home with us a large fee, which our respective hospitals allowed us to keep. As far as my family was concerned, you can almost guess the rest, especially bearing in mind Rosemary's chosen career. Yes! - in !962 the whole family set out for a fortnight's holiday in Austria – a week in a villa in the mountains above a little village called Zell-am-See – followed by a week in Vienna. The week in Zell seemed almost literally 'out of this world' – with towering snow-capped mountains as the backdrop, and our itinerary from the villa to the lake taking us through alpine meadows comprised of a veritable carpet of alpine flowers. But our stay was marred by a near-fatal incident involving Roger.

Not much of a swimmer myself, I had taught the children to swim on their backs, partly because of the simple change from swimming to floating, and back again. The two boys and I were in the margins of the lake, nicely 'within our depth, and Roger went past me, on his back, as I had taught him. Then, instead of just floating for a bit of a rest, he decided to stand up. We had not, however, been warned that there were sudden and serious changes in depth, associated with rocky ledges under water, and Roger, instead of being able to stand, found himself out of his depth, and unable to get himself into a swimming position again. The next thing I knew was that he was bouncing up and down in the water, and barking like a dog every time he bounced upwards. What I didn't realize at first was that he was gasping for air before going under again.

Then, suddenly the awful truth dawned on me, and I swam the few strokes to him, and (as the evidence showed all too plainly afterwards) clawed frantically at his back in the endeavour to get him floating again. Some way or other which I have never understood, I managed to move him along by the few feet necessary for him to be able to stand again, and get his breath back.

We agreed not to tell Mary what had happened, lest it frightened her unduly (she was waiting on the shore with the towels), but there was something we hadn't reckoned on: great long, red weals all over his back, left behind by my finger nails, as I had clawed at him so desperately, fearing that he was going to drown.

Mary exclaimed, 'Good Heavens! What are all those marks on your back?' Roger didn't even know they were there – nor did John and I initially, and somehow or other we managed to conceal their real origin from Mary. The horror of those few moments is with me still.

We moved to a hotel in Vienna for the second week, and what a wonderful time we had there, too! The whole city is steeped in musical history, commemorated by statues to great composers such as Mozart, Haydn, Beethoven, Johan Strauss, Brahms and Mahler. Beethoven's association with the city ran particularly long and deep. He moved to Vienna in 1792 and lived there for the remaining 35 years of his life. He was, in his way of life, a very restless man apparently, and is said to have lived in 60 different places in the city and its environs during those 35 years. How well do I remember a day spent in the Vienna Woods! One would hardly have batted an eyelid to meet Beethoven himself at any moment – with shoulders rounded, head down, and that familiar scowl of concentration spread over his features. There was the evening meal, too, that we partook of in the famous wine cellar, whose domed and plastered ceiling was covered with the signatures of famous people, including that of Franz Schubert. The meal was memorable also for the effect that a few gulps from a glass of Viennese red wine had on Rosemary at the tender age of twelve: she was distinctly merry after it, and in need of some support on the walk back to our hotel. As in my spell in the city the year before, the highlight of our stay in the city, needless to say, were visits to the Musikverein. What a holiday! – and, in fact, it was to turn out to be the only holiday we ever spent as a family, abroad.

On my way back in the plane following my professional visit in 1961, I wrote a poem specially for Rosemary. Here it is.

Making Music

Let scientist dissect this
symphony of sound, evoked by
men, full hundred, from instruments,
cunningly devised, long since, by
those who, knowing nought of science,
yet had the instinct and intuition to
fashion from prime matter –
wood, skin, hair, and shining metal –
the mouthpieces of heaven.

Let him transcribe them
into waveforms – display, and
analyse them as spectral distributions,
in the only language that he knows;
as well, let him dispense with such
archaic crudities as violins, and flute's
elusive voice, and substitute mere
sinusoidal oscillations – fundamental
with harmonics mix – match
note for note, and tone for tone,
these relics of the past.
Might it not be argued
that in such terms lies
deeper understanding
of the composer's art?

> Yet would it lack
> the one ingredient needful to
> add Eternity to music's voice –
> for music needs men to make it.
> And let it be with horse hair
> given bite by dried sap,
> drawn across twisted
> sheep's gut with the touch
> of God; with pursed lips, too,
> coaxing, God knows how, from
> flute and flageolet, from
> trumpet and trombone,
> the soft sibilants of woodwind
> and triumphal tones of brass;
> men who, agonized by the notes
> before them, suffer the pangs
> of childbirth to create anew
> the masterpiece,
> each time they play.
>
> > This is the fashion in which
> > music is ever made, and
> > ever will be – by men who,
> > having heard the voice of God,
> > cannot keep silence.

Since I wrote that, 30,000 feet up over Europe, on a dark winter's night all those years ago, I have become more and more convinced that, second only to pure laughter, music is indeed the language of Heaven.

Chapter 14

From the perspective of fifty years on, I believe it would be true to say that the middle 1950s represented the peak of my career at The London Hospital. It was marked by the paper I read in Copenhagen in 1954, on errors in the British and American X-ray dose standards. acknowledged by the award of the Roentgen Prize of The British Institute of Radiology in 1956.

The ten years that followed, before the move to The National Physical Laboratory in 1966, concluded with a period of considerable interest and activity in the field of solid-state electronics when it was developing very rapidly, but before the era of integrated circuitry and 'micro-chips' had arrived. As an early exercise in the art, I developed a comprehensive piece of equipment which displayed (on a cathode-ray tube screen) a complete set of 'characteristic curves' of any transistor plugged in to it. Every single component of this piece of equipment was 'solid-state', except the cathode-ray tube itself! A number of other instruments followed, including transistorised door-post alarm systems to detect unauthorized, or accidental transport of radioactive materials, as well as completely solid-state X-ray treatment monitors incorporating a number of novel features. It was a productive decade but, at times, it was anything but a happy one, either professionally or domestically.

One night In 1961, as Mary was going to bed, she had a terrifyingly severe nose bleed. We immediately associated it with high blood pressure, though we were told subsequently that there was no direct connection between the two. Nevertheless, a visit to the doctor revealed that she did in fact have very high blood pressure. It was sudden, and unexpected news, that seemed to change the whole landscape of our family life, and of life generally. Until then, during the twenty and more years that I had known Mary, she had seemed so fit (rarely succumbing even to a cold) that I had been lulled into thinking that she would remain so, until old age finally began to take its toll of both of us. Alas, that was to be far from the case.

But, professional matters first. My work on the discrepancies between the British and American X-ray dose standards had shown the errors to be due to distortion of the electrostatic field between the electrode plates. The electrode system in each so-called 'free-air' ionisation chamber consisted basically of two flat parallel metal plates, and the electrostatic field lines between them should consist of straight lines perpendicular to both plates. However, the (albeit necessary) metal box housing the electrode system will distort the field (unless steps are taken to prevent this by 'guarding' the field from these effects), resulting in the errors observed. To prevent this from happening, I incorporated a system of equal-width metal strips along each side of the chamber, with narrow gaps between them, and equal increments of potential between adjacent strips. This replaced the system in use at the time in the international standards, of equally-spaced wires, and was much more effective in straightening the electric field lines, making sure that all the electric charges released by the radiation beam were collected and measured, thus eliminating the errors.

Before I moved on to radiological applications of solid-state circuitry, my researches during the second half of the nineteen-fifties were taken up mainly to applying the principles of field-guarding to a graphite-walled 'thimble' chamber, which, as the name implies, is compact, and encloses a cylindrically-shaped air volume. By constructing the cylindrical walls of equal-thickness graphite rings insulated from one another (by thin ,plastic-film annuli), and (as with the guard wires) by arranging for equal increments of potential to be applied to them, the electrostatic field in the cylindrical space could again be straightened up, thus enabling a truly cylindrical 'collecting' space to be defined (a little smaller in diameter than the actual cavity), whose volume could then be calculated simply and accurately – information necessary for the 'absolute' determination of radiation dose which, in this case, would be high-energy radiation such as that provided by kilocurie radioactive cobalt Units.

The theoretical determination of the electrostatic field configuration in such a complex system was undertaken by my young colleague Barry Barber (who had joined the Department straight from Cambridge) as the basis of a PhD thesis. (Incidentally, in recent years, after a very long interval, we have become re-acquainted, with an inevitable exchange of reminiscences about 'the old days'!) In fact (and I suppose it stems from its being his first job out of college) Barry has, during these reminiscences, revivified for me much of the lost detail of those productive and exciting years.

But to return to my story: the objective of this work was to achieve a design of graphite cavity chamber which could be used as a dosimetric standard for high-energy radiation, analogous to the 'free-air' chamber standards used for low and medium energies, and its successful outcome was described in a paper I read at the International Radiology Congress in Montreal, in 1962. It is here, however, that the story becomes somewhat convoluted, with - sad to say - a dark twist to it which, eventually, was to change the course of my professional life. Although, at the time, the direction that events took seemed almost totally disastrous, nevertheless, from the perspective of nearly fifty years on, I am able to see the whole sequence of events as a profound example of the wisdom embodied in the Chinese character symbolizing 'disaster' – including, as it apparently does, the symbol for 'opportunity', as well. The events which I am about to describe spelt 'melt-down' for me, but it was a 'melt-down' that, so to say, enabled me to be 'poured out' of the melting pot into a new mould, able to contemplate, and embark on, an entirely new phase in my professional life. Moreover – though, of course, I was not aware of it at the time – it was preparing me, both mentally and emotionally, for the task which lay only a few years ahead – that of caring for an invalid Mary, throughout the last twenty years of her life.

The paper I read at the Montreal Congress described a

definitive experiment which Barry and I had carried out to measure the output of The London Hospital Kilocurie Cobalt Unit which we had installed, following the tour of Canada and The States by Dr Boden and myself, in 1954. The measurement was made, using the guarded-field graphite thimble chamber as the reference standard, and agreed to within a fraction of a per cent with the output as measured by the use of a portable so-called 'secondary' standard', calibrated by The National Physical Laboratory. So far, so good. How, then, could my account to the Congress of such a successful outcome have sown the seeds of disaster for me? You could say (as a former acquaintance of mine was fond of doing in such an event), that it was merely due to 'a concatenation of unfortunate circumstances'. As such, I could hardly have anticipated the consequence for me of my trip to Montreal - and perhaps that was just as well. I shall do my best (and, with the benefit of hindsight) to string together, as logically as I can, the whole sequence of events following the delivery of my paper to the Congress, and to show how, in a sense, 'one thing led to another' - despite the fact that they would seem to be unrelated - with the psychological pressure on me building inexorably, until it reached my breaking point.

  Actually, the first event in the whole sequence could be said to have resulted from what might be described as a 'scientific whim' of mine: I loved working at the level of what one might call 'first principles' – particularly in the field of standardization. The measurement of the output of the kilocurie cobalt unit consisted, fundamentally, and quite simply, of the measurement, as accurately as possible, of a very small electric current. So - we carried it out by means of the simplest and most basic pieces pf equipment that could be employed, namely a moving coil galvanometer, a set of standard resistance boxes, and a so-called 'standard' voltaic cell, providing a very accurately-known current which, by means of the galvanometer, could be matched to the current produced by the guarded-field chamber, using a so-called 'null' method. It couldn't have been simpler, or more fundamental, as a measuring technique. I suppose there was a certain, almost 'aesthetic' appeal to it – a certain elegance in reducing the instrumentation employed to its most fundamental, and – yes – its most obviously 'trustworthy' format. There is no doubt that these personal preferences of mine were further entrenched in my mind by the nature of the whole project – that of the absolute measurement of radiation dose, which, to my way of thinking at the time, called for reference to other standards, e.g. of resistance and potential. And there is no doubt also that this would have caused me to give less consideration than I might otherwise have done to legitimate alternatives. Furthermore, the innovation at the heart of the paper which I was presenting was the guarded-field graphite thimble chamber: the method I chose to use for the measurement of the current it produced was of no real consequence, or, indeed, relevance.

Despite this, one of my audience chose to ignore the fundamental advance which the design of the chamber represented (and the highly satisfactory outcome of the whole research), and asked me why I hadn't used a so-called operational amplifier to multiply up the current produced by the thimble chamber before measuring it. Fatefully, I was unable, on the spot, to convey to him the subtle personal feelings and preferences that lay behind the *modus operandi* I had chosen, the inference made then (but not brought into the open until later in my stay in Montreal) - that I had 'fallen behind the times' in my working practices... Who knows? – perhaps I had, in the matter of how I had chosen to measure the chamber current – as I have said, of no real relevance to the innovative advance at the heart of my paper.

'Until later in my stay' was, in fact, a weekend party at the lakeside cottage of one Professor Harold Johns, whose notions regarding the entertainment of his guests, like most others of his ilk in North America, had a pretty 'rough and tumble' side to them. As an example (and before pursuing the main theme of my story), regardless of one's swimming abilities (or lack of them) we were each expected to make an attempt at water skiing, which involved being towed behind a speedboat (yes, we were allowed a life jacket) and try our hand (actually our feet) at getting up onto the skis when a sufficient speed had been reached. In my case – barely able to swim ten yards – it was an absolutely terrifying experience in which, frightened to let go of the rope until circumstances forced me to, I was towed along, at speed, under water... When I did let go, I was left alone, with the flimsiest of life belts, barely afloat, in the middle of the lake, whilst the speedboat, having proceeded several hundreds of yards further, finally turned round and came back to pick me up. It was all regarded as the occasion for a good laugh, whilst I, inwardly, felt I was retracing my steps from death's door. Little did I know that it was the forerunner of some much subtler, and, potentially, much more (mentally) damaging games, to come.

The occasion was the party that followed dinner that evening, and after several other traditional games, we had reached 'Charades'. And fatefully, some Clever Dicks in the company had homed onto the word 'galvanometer', and its individual syllables – 'gal', 'van', and 'Oh, meet her!' It was shock enough, to find my perfectly tenable 'first principles' approach had become the subject of cheap jibes. But it got worse. The cottage was 'open-plan', with a staircase leading up from the lounge to a railed landing. One of the party took up a position there, leaning over the rails, and seductively gesturing 'Come up and see me!' - to the great amusement of the assembled company - with the exception of myself, of course, my heart already in my boots. 'Oh, meet her!' the rabble cried. I can't remember how 'van' was fitted into the story line. But does it matter? I had flown out to Montreal feeling myself to be 'one of them' – to wit, the whole band of radiation physicists dedicated to improve the use of radiation in the treatment of cancer: what a privilege! But I flew home again, feeling myself to have been made fun of by my fellow scientists, none of whom, apparently, shared even a fragment of my sense of satisfaction at making a fundamental measurement by employing the most fundamental of methods: I felt I was no longer 'one of them.'

Back home again, I couldn't even bring myself to tell Mary what had happened. My perspective had been destroyed, and I found that I myself – yes – I myself was beginning to ridicule the manner in which I had chosen to make the measurements in order to satisfy what I was now regarding as nothing more than a whim of (what one might call) 'scientific aesthetics'. My belief in myself as a successful, innovative medical physicist had by then virtually gone out of the window.

What next? Before answering that question, it will be helpful to recall the answer to another question: why did I find myself in Montreal in the first place? Surely, the answer couldn't be simpler: to report to the Congress the satisfactory outcome of the work Barry and I had carried out on the use of the guarded-field thimble chamber for standardization purposes. Whilst that is true, a complete answer goes back further than that – much further – and reminding ourselves of that will be helpful in understanding what did actually happen next.

It goes back, in fact, to my childhood, and the extreme fear of cancer from which my mother suffered. Deep in my subconscious, this did, indeed, ensure that, sooner of later, I would have to have my own reckoning with the disease – not necessarily in my body, but in mind and spirit – and not for myself alone, but for and on behalf of others. Fired up thus in 1944 for my personal crusade - not against Hitler, but against cancer - I embarked on a mission which was to produce the radium dose calculators, the automatic X-ray dose plotter, the laying bare of discrepancies in the International standards of dose, the guarded-field thimble chamber, and much other research and development, stemming from these projects. Far from being 'all in a day's work', all this was undertaken with a zeal matching that of any front-line soldier, and it was taking its toll on me. In the background, and ever haunting me, was the fear that my department might make a mistake in its measurements (or calculations), resulting in the deaths of dozens of patients - mistakes which might not show up for weeks or even months, when it would be far too late to rectify the errors.

It was this sombre back-drop to one's working day which, if one was to get on with one's work and life, one had to do one's best to ignore. Nevertheless, it was at no time entirely absent from one's mind. I suppose that, despite this, my confidence in myself - in relation both to my work and the responsibilities it involved – had grown steadily, over the 16 years that had preceded Montreal, although - because of experiences with my mother in early childhood - I had never been able to approach the work with the day-to-day matter-of-factness of the majority of my colleagues, including those who had, all unwittingly, dealt me such a blow whilst I was there.

So, after Montreal, it was back to work as usual. Or was it? Alas, no. From the first moment, I was aware of the subtle change in myself. Small decisions begin to feel like big ones, and instead of feeling I was one of a highly supportive team, I began to feel isolated from my colleagues ('You're on your own now,' I seemed to be saying to myself), and, fatefully, I found myself continually pricking up my ears for any further evidence of criticism. *Without realizing it, I was becoming paranoic.* And looking back, once more with the benefit of hindsight, I realize that a devilish feature of this condition is a greatly enhanced propensity to mis-hear – no, not so much to *mis*-hear, as to twist what one has heard, into what I was more than half fearing - or expecting - to hear.

The die was cast, and it was only a matter of time before I was to think I had heard something confirming that what I had feared all along had at last occurred: a devastating radiation accident, for which my department had been responsible – I myself having the ultimate responsibility. And as if that weren't enough, my fevered spirit gave the whole thing a final fiendish twist in the tail – the accident had been so ghastly, I seem to have gathered, that no-one had dared tell me about it.

The rot set in. In almost every conversation or discussion I had with colleagues, I felt sure I was detecting sly references to the disaster that was being kept from me, and it wasn't long before I felt that, for the time being at any rate, I could no longer fulfil my role as head of the department, and I fell into the hands of the psychiatrists, and even spent a brief time in The Maudsley Hospital. It was some months before I was able to escape from my tormented state. It had been a desperately traumatic time, not only for myself, but for Mary also, and the children.

During those months away from the Department my colleague Barry was acting Head, and, for both of us, in our separate ways, it proved to be a definitive time – Barry subsequently moving on to ground-breaking research of his own initiation – nothing less than pioneering the use of computers in the Health Service, a mission for which he was so well equipped. I am in no position to give an adequate account of this work, or to convey to the reader the scale of the impact which it eventually had within the whole of the Health Service. What I *can* say, however, is that it all began with our becoming aware, among other things, of the peaceful applications of the mathematics embodied in so-called Operations Research – originally developed as a means – yes - of enabling men to kill more effectively: 'swords into ploughshares', one might say. We took ourselves off to a course of lectures on Operational Research, after which, figuratively, I doffed my hat to Barry, and, in effect, said to him, 'You're on your own now – go for it!' And go for it he certainly did! – as exemplified by the mountain of papers and reports which followed, over the years.

In February 1966 an Operational Research Unit was established at the Hospital, with Barry as its Director. Its interests would cover an enormously wide range of administrative procedures, from the most efficient daily deployment of a reserve pool of nursing staff, to the application of so-called 'queueing theory' to the organization of attendance at out-patient clinics, to say nothing of a rapidly-growing involvement with the practices and procedures employed in the administration of the finances of the Hospital, not the least of which was the transition from punched-card systems to computers – in those days very much in their infancy. And that was only the beginning! The history of the expansion of the work, involving the introduction of computers into the whole of the National Health Service, would fill several volumes.

But before leaving the subject of Operations Research and all its ramifications, I can't resist including a trenchant little poem. The original version was in fact actually *called* 'Operations Research', but was lost. However, the form and substance of the handful of deceptively simple little verses which follow, parallel those of the original poem, ('Where "y" stands for "yield" – or men in battle killed', as the poem had it). This later version was written in the 1980s, on hearing the then Chancellor of the Exchequer counter the bankruptcies' statistics merely by setting them against the statistics of the number of new businesses being created.

The Advantage of Abstraction

'In algebra', the teacher said,
'we deal with things much better,
by simply placing in their stead
an algebraic letter.

'By this device, with power imbued,
we are in fact enabled
to calculate the magnitude
of quantities so labelled.

'For businesses new-formed an "n",
for bankruptcies a "b";
should "n" but "b" exceed - why, then! -
forget the misery!

'The beauty of the method's plain,
for when with symbols dealing,
a "b" 's a "b", and it is vain
to think that it has feeling.'

Barry, then, was moving on, and for my part, as I have already implied, I had become convinced that I had served my time on the clinical side of radiation physics, and, in due course, would need to move on, too. It was not a negative decision: it felt the right move, at the right time. And as if to substantiate that conviction, in 1966, Dr Aston, Head of the Low and Medium Energy Dosimetry Group at The National Physical Laboratory retired, and I was head-hunted for the job! But I anticipate.

I would be failing in my task if I were to continue my story without recounting something which was crucial to my full recovery, and which in due course enabled me to make that move to an entirely new phase of my career. How well I remember the occasion when I spoke of it to one of the psychiatrists with whom I had been involved at The Maudsley Hospital!
I said, quite simply, that as an outcome of all that I had suffered, I seemed to have stumbled on something which I referred to (almost skittishly) as 'the non-imputing principle'. It did, literally, enable me to shed the last remnants of paranoia, and it has, in fact, stood me in good stead subsequently, whenever I have felt tempted to impute anything to another, be he or she colleague, friend, or perfect stranger. I spelt it out in some such words as these: 'If someone — anybody — is maligning you, in whatever way, then the only person it can really harm is themselves. But if you *impute* something to them — it will most certainly harm *you*.'
After nearly fifty years, in my mind's eye I can still see the expression on the psychiatrist's face as I was speaking. Unlike the voice, which can say only one thing at a time, his expression seemed to be saying two or three things simultaneously. One was being expressed by the smile that spread across it that seemed to be saying 'You've *done* it! You're out of the wood at last!' But at the same time his eyes seemed to be scrutinizing me, looking for some clue as to what it was that had enabled me to arrive at such a conclusion, and to express it in such

elemental terms. And somehow, and in whatever way (I don't know quite how, even to this day), he conveyed the distinct impression that I had managed to put into a form of words — words with which he was not familiar - an important psychological truth, AND that he had got as near as he was likely to, to saying, 'Thank you for that.'

Chapter 15

The events of Chapter 14 were soon to be followed by the severe winter of '62 – '63, with heavy snowfalls and sub-zero temperatures arriving before Christmas, and persisting until Easter – a winter unlikely to be forgotten by anyone who experienced it. But why mention it here? The fact is, that like thousands of other families across the country, it led to tragedy in ours.

It will be recollected that my parents had come to live in the next road to ours nearly ten years before, and from the beginning they were in need of some help and support from us – not in the financial sense, but logistically, in the matter of heavy shopping, and the likes. So it was that once Christmas was over, and for reasons I cannot recall in detail, we were not only involved in extra shopping expeditions on their behalf, but soon found ourselves taking round hot meals for them, through the snow. We all thought that it would be for a week or two, at the most. But, as the weeks went by, with the weather showing no signs of relenting, we came to feel that they would be safer, and better cared for *pro tem*, in a home for the elderly. It would not, could not, be for long, we assured ourselves – the snow couldn't possibly last that much longer. So it came about that they were taken into the care of a Home in Farnham, a few miles west of Guildford: there was a good bus service between the two towns, so there would be no difficulty in making regular visits.

Alas, in the event, the time scale came to be measured in months, not weeks, and what began as a satisfactory arrangement for a week or two, was far from so, as the weeks became near-unbelievable months. Not uncommonly in those days, the *modus operandi* of the Home was based on separate quarters for males and females – even for those who were married. As I have said, this would have been acceptable for a week or two, but when weeks became months, both my mother and my father understandably began to fret. Nevertheless, we were at the mercy of the weather, and it was the weather that took final toll of my mother. She went down with a bad cold, which rapidly turned to pneumonia, and soon she had died. It was a shattering blow to us all – particularly to my father – and totally unanticipated.

When my father left the Home, soon after my mother's death and funeral, we had to find accommodation for him with a caring widow, who lived alone a mile or so away: our small, cottage-style house, with only three bedrooms, could not accommodate him, as well as our three children. But months later, when Roger had left home for Liverpool University (John was already up at Manchester University), my father was able to come home to us. Understandably, a year or two only from his Golden Wedding, and finding himself without his partner of so many years, he was a far from happy man, and slowly deteriorated both physically and mentally, and he, too, died a year later, in 1964.

The events of Montreal, with their awful consequences for me, coupled with the loss of both my parents during the following two years, certainly made the earlier part of the sixties a very sombre time. But in 1964 and '65, Mary and I began to find our feet again. I was back in full harness in my Department at The London Hospital, and Mary, as resilient as ever, went back to school (so to say!) at Guildford Polytechnic. There, she embarked on further A-level courses new to her, and which would enable her to go on to study for the London University Extramural Diploma in Social Studies: she hoped to do case work in due course. We seemed to be emerging from the deep, dark shadows of the early part of the decade, and to be entering an entirely new phase of our lives. By 1966 Mary had acquired her requisite A-levels with good credits, and I had started work in my new post at The National Physical Laboratory at Teddington, as Head of the Low and Medium Energy X-ray Dosimetry Group. One might say that the new post fitted me like a glove! My researches at The London Hospital into the discrepancies between the international standards made the move seem the most natural one in the world at that juncture, with the bonus of being able to leave behind me the day-to-day responsibilities for the radiation doses administered to the constant stream of patients passing through the Radiotherapy Department, and which, combined with the loss of confidence created by my experiences in Montreal, had finally broken me.

To mark my departure, I took the whole of my staff, together with the consultant radiotherapist, Dr. Walter Shanks, to a meal out, and a concert at The Royal Festival Hall, on the South Bank. It was, indeed, a memorable occasion, and well do I recall – as though it were only yesterday – going with Dr Shanks (during the interval) onto the balcony overlooking the Thames, and exchanging deeply personal reminiscences. One very unusual one I remember so well – of Dr Shanks, awash with the wonderful music to which we had been treated during the first half of the concert, confessing to me that, despite his medical career, the one thing he had always aspired to, was to write a novel, whose title would have been 'The Symphony and the Crowd'. He didn't expand on the title, and, I suppose, considering the occasion, the matter was well left. But thirty years later it was I that found myself writing an (as yet unpublished!) novel, called 'Michael's Notes', about a brilliant young symphonic composer (the nearest I would get to realizing a persistent childhood dream of mine!). Alas, Dr. Shanks had recently died, and I was unable to share it with him – to 'compare notes', as you might say... At this point, it is worth recording that within a few days of taking up my new post I had received a letter from him, thanking me for all the work I had been responsible for at the Hospital, and concluding with the statement that during my spell of twenty years in charge of the dosimetric side of the treatments, 20,000 patients had passed through the Department, without a single radiation accident or mishap of any kind. I still have that letter – the means whereby I had been able finally to lay to rest the nightmare of Montreal.

It was to be expected that I would feel that my new post also involved heavy responsibilities, but they were much more academic in nature, and I had left behind me the day-to-day clinical involvement with a roomful of cancer patients. Moreover (thankfully) I had no thought or feelings that I was 'abandoning ship', so to say. I was simply moving to the more academic, and scientific side of things, but I would still be serving in the war against cancer, which was still so important to me.

Happily, too, my earliest experiences in my new post had their amusing side. One of my new colleagues was one Andy Marsh, a lovely, friendly, guileless man, who was soon teasing me about my reputation having 'gone before me'. In a facetiously conspiratorial voice, he informed me that when they heard I was coming, they hid much of their equipment under the bench - well out of sight! His joke was a good omen, and an augury of what the remaining twelve years of my working life were to be like — namely, that they were to be spent with as nice a bunch of colleagues as one could wish to have. Little did I know, however, that within two years of my taking up my new post, Mary and I were to be plunged into another crisis that would continue, in one form or another, until Mary's death - twenty years later - in 1988.

One might say that the three years, 1965 to 1967, provided the lull before the storm that broke in 1968, but they provided much more than that: looking back on them with the perspective provided by the intervening 40 years, one might almost say that they were halcyon years. As I have said, Mary and I certainly found our feet again during that time, and both of us seem to be positively rejuvenated, as changing circumstances opened our lives up to the influence of new friends and fresh associations. One such new friendship formed during this time turned out to be of great significance to me. It will be recollected that it was during this time that I had to resign my membership of the Quakers (remember? — I quaked too much...) and we started attending All Saints' Anglican Church, situated in the heart of Onslow Village, with Tom New its young vicar, of just a few years' standing in the post. He was a man after my own heart, and we quickly became close friends.

It was Tom who introduced us to 'Clinical Theology', being pioneered at that time by Frank Lake. As the name implies, it was a movement led by informed clergy such as Tom, whose aim was to produce what might be called 'lay' counsellors with skills consisting of the basics of clinical psychology, illuminated and informed by a spiritual approach to them. It was a fast-growing movement at the time, but, sadly, in 1968 (as chronicled below) we had to end our involvement – which in Mary's case had already become a practical one, as a counsellor. A synopsis of Frank Lake's pioneering work is to be found in Appendix 2. Realizing that I had no information to offer on the subsequent history of the movement which Frank Lake had initiated, I typed 'Clinical Theology' into 'Google', to find immediately that it has certainly survived, but is now known as the 'Bridge Pastoral Foundation'.

It was through Tom, too, that I learned of a group based on Guildford Cathedral (less than a mile away), known as 'The Guildford Cathedral Religion and Science Group'. It was, in essence, a 'learned society', small in numbers (under 20), but making up for its smallness by the great enthusiasm with which it pursued its business. Which was? – in a nutshell, nothing more nor less than to explore the relationship between the scientists' approach, and what, for want of a better word, we will call the theologians' approach to human existence, and Man's place and role in the Universe in which he finds himself. Not quite the nutshell I promised, but I wanted to make sure that it is understood that, small though the group was, its remit was of the broadest and deepest. Its proceedings were as formal as any such group: we took it in turn to read papers before the group, followed by a discussion, conducted equally formally. There were scientists from the University (hard by the Cathedral), and a sizeable contingent drawn from the Cathedral staff. And there were also 'unattached' members, like myself. One of these was one Martin Israel – pathologist-cum-priest, and the author of a number of significant theological tomes, one of the best known of which was 'The Pain that Heals', my copy of which is the worse for the wear – the result of frequent consultation.

My involvement in the Group was a daunting experience at first. I had long since been at ease reading a scientific paper to a bevy of fellow scientists, but to read what amounted to an 'amateur' submission on theology and/or spirituality to a gathering of Cathedral dignitaries and University experts was quite another thing! We were a close-knit group however – intimate, it would have to be said, rather than formal. For example, at one point I received a hand-written letter from Martin Israel expressing his appreciation of what I had said in one of the papers I had read. There were, in fact, three of these papers altogether, entitled 'Has Science so Changed Life as to Make Religious Belief More Difficult?', 'The Nature and Role of the Non-rational in Human Life', and 'The Scientific Approach to the Resurrection.' Almost miraculously, all three have survived, though the copy I have of the third one is subtitled 'Part 2'. As I have no Part 1, I suspect that the subject itself was divided between two evenings, my contribution being on the second evening.

These three contributions are reproduced in Appendix 3, for those who might be interested. The third one was subsequently suitably edited and put on tape, and, with a new title, 'On the Third Day...', was submitted to the BBC as a quarter-hour Passion Week talk. They did not make use of it at the time (it is some ten years, at least, since I submitted it), and, as I write (January, 2008), I am thinking that I might try again, in time for Easter. Times do change, and, recently, what I call the 'evangelistic scientific atheists' have overplayed their hand somewhat. Who knows? – the time may be ripe now, for what I had to say all of forty years ago...

I recollect one other commitment during those years of new beginnings in the mid-1960s, and one which is worth more than a mere mention. I undertook it in anticipation of the move to The National Physical Laboratory, and - commenced before I had actually started - it was completed during my first two years in the new post. The time and place for all the hard work involved was, in fact, confined to my journeys to and from work (how much I owe to what was then known as 'The Southern Railway'!). I had been aware of the existence of a certain book which had been published by McGraw-Hill in 1958 – it was by one Louis A Pipes, and was called 'Applied Mathematics for Engineers and Physicists' (perhaps a better title would have been 'Mathematics in Engineering and Physics'). Be that as it may, awareness of the book, and the reputation it had earned for itself, had been gnawing away at me for some years: I had long felt both its challenge, and its potential in providing me with a new and powerful tool in my work, particularly in meeting the demands which I might face in fulfilling my duties at the NPL. So – for over two years I utilized the time I spent commuting (approaching two hours a day) working my way through all 713 pages of the book, including every single problem! – and there must have been several hundred, altogether. Up until that time, my mathematical knowledge had consisted of what I'd learned as the 'subsidiary' subject in my honours Physics degree – not an insignificant amount, by any means, but, on the other hand, not inclusive of the likes of matrices, and such truly beautiful and powerful mathematics known as Laplace and Fourier Transfoms. As things turned out, my last major project at the NPL, before retiring, would have been impossible, without the knowledge I acquired almost incidentally, as I travelled back and forth, to work. There is a proverb (possibly Chinese) which goes something like this: 'If you add a little to a little often enough it grows into something big'. It may seem to be stating the obvious, but there - sometimes the obvious is worth stating succinctly. For myself, I've never been well disposed towards 'the Herculean effort', but I do seem to have been rather good at the business of persisting in adding a little to a little...

Chapter 16

There came a fine June morning in 1968, when I left home as usual, to jog-trot my way down to the station to catch the train to Surbiton, and thence, by bus, to Teddington – just one more of those commuting days. The earlier, fraught years of the 60s were well and truly behind us, and life for both of us seemed set fair again.

As I mounted the steps of a little snicket that led into the next road (and my shortest way down to the station) I turned, as usual, to wave to Mary. How was I to know, as she waved back, that that would be the last I would see of Mary as I had known her for thirty years? As the train sped towards Surbiton, my mind and heart were filled with joy, as I contemplated the new phase our lives were entering; and, by the time the train had reached Surbiton, (having abandoned the mathematics for once) I had expressed the joy in a brief but heartfelt poem.

I just couldn't wait to share it with Mary, and I rang her immediately I got to work: her day, as well as mine, then, would be suffused with its happy anticipation of our new future together. I got no reply, and repeatedly so, as I rang at intervals throughout the day. An insidious fear began to eat away at the joy with which the day had begun. I tried to console myself with the thought that Mary had gone off early to spend the day in the library, in the furtherance of her studies, but the thought carried little or no conviction, and when I rang from Guildford station on my way home, and there was still no reply, I knew that disaster, in some form or other, was facing me.

An unclaimed bottle of milk greeted me on the back doorstep, the unlocked door opened to a turn of the knob, and there was no response as I called out Mary's name, and no sign of her on the ground floor. I mounted the stairs two at a time, in real panic now. And there I found her, unconscious on the landing, in a pool of vomit. It emerged, subsequently, that she had gone upstairs to make the bed but had suffered a severe stroke before she had even reached the bedroom, and had lain there all day, from shortly after I had left her in the morning.

As I bent over her unconscious form, I was clutching the poem I had wanted so much to share with her. It was many weeks, however, before I would be able to do so, and even then, the Mary I eventually came to share it with was not the Mary for whom it had been written. In her very response to the suffering she had undergone , she had, perhaps (quoting T S Elliot), been rewarded with ' a tremor of bliss, a wink of Heaven, a whisper...' – yes – a 'whisper' that had given her response to the affairs of this life a certain 'otherworldliness', which meant (with such sadness for me) that I would never know what the response of the Mary to whom I thought I was coming home would have been, had she been there as usual, to welcome me.

As I subsequently wrote, in a little poem,

> Sometimes
> I have thought
> that on that day
> in June, what had
> just been given me –
> a Morning Vision
> of Evening –
> was taken away
> again immediately:
> for darkness had fallen
> before evening
> had begun; or so
> it seemed to me.

And the vision? – it looked back over such times as those which had followed that fateful visit to Montreal, and forward to the future which it seemed we had won for ourselves, after 'the long day's heat'.

> What does it matter,
> but that I love thee? –
> with being melted
> in the crucible of past
> tormented meetings,
> and re-cast
> in the desolation
> of failure
> to meet with thee
> at all.
> What does it matter,
> but that I meet thee now? –
> my being with yours,
> finding each other,
> and ourselves, anew,
> as the steaming mists
> clear, after the long day's
> heat, and early evening's
> cool refreshment renews
> the morning's promise;
> and we are free
> to love in an eternity
> of time, deprived,
> once for all,
> of the tyranny
> of its passing.
> What does it matter
> now, save that
> I love thee?

Even as I type those words from the original hand-written copy on the (now yellowing) sheet of foolscap I had had with me on that fateful journey to work - even as I type them, the contrast between my vision for that evening, and what actually transpired, is as stark – and, yes, as *mystical* - as it had seemed when I dropped to my knees beside the unconscious Mary, forty years ago. Twenty years on, in 1988, I had re-read the poem, in the context of Mary's death – and it seemed to acquire a still deeper meaning and significance.

Were I to dare to use the word 'fortunate' about any aspect of such an event, it would have to be that it was indeed fortunate that it hadn't happened two years earlier. Then, just a few weeks before I was due to start my new job, or – worse still – a few weeks after I had started it, it would almost certainly have meant the end of my new career before it had even begun. As it was, I had had two years in which to establish myself in it, and to prove to myself that I was 'up to it'.

And – to my great blessing - when the blow fell I discovered that my colleagues were not only very proficient physicists, but wonderfully supportive human beings as well. And it didn't end there: there was something different about them as scientists, something which I had become aware of almost subconsciously during those first two years, but which was becoming clearer to me now, under the pressure of events. I realized that, whilst applying themselves to their work as enthusiastically as the next man, they were, so to say, more available, as human beings (unlike many of their fellow scientists, who prefer to deal with the problems their science sets them, rather than the problems that life poses).

Why so? – I wondered. It occurred to me then that, rather than their being out to make a name for themselves in academia, or to make a lot of money for an industrial employer, their basic role in life was, as one might say, that of the custodians of standards of excellence in measurement - from the physical standard of length, to the atomically–based standard of time. That's right! – they were not part of the scientific rat race, in all its various forms, which meant that they had remained more 'rounded' as human beings; and – yes – who, as such, immediately rallied round me, in the deep trouble that had beset me. That may not be the whole story as to why I felt my NPL colleagues to be so supportive – I'm sure it's not. At least in part, it was simply because (as I have said earlier) they were 'as nice a bunch of colleagues as one could wish to have'.

So it was that, with their understanding and support, and the understanding and support of our friend Tom New, and other friends in the close-knit community of Onslow Village,

Mary and I weathered the worst of the storm, through the late summer months of 1968, and into the early autumn. It hardly needs to be said, that expressing my feelings at such a time in poetry was also a great help to me. And here is just one of the poems in which I endeavoured to describe the nature and extent of the change that had come over our lives, so catastrophically suddenly:

> The year wears on,
> the magic mists appear,
> casting their immemorial spell;
> and leaves,
> fresh green when you were well,
> are turning brown, then red,
> and, twisting in the chill wind, fall
> as dead—
> trees shedding
> their sad confetti.
>
> Daily, the artist sun,
> with prodigal palette,
> paints cosmic canvases;
> and, night by night,
> the stark stars,
> piercing the canopy of evening,
> stave off, still,
> the gathering dark.
> and birds, on branch and eave,
> incredulously yet sing,
> to catch my spirit
> off-guard,
> and evoke
> the fierce pang
> of remembered joy—
> joy that I scarce now
> dare contemplate.

Considering that it was Mary's first experience of serious illness, it is amazing, looking back, to realize the extent to which she took it all 'in her stride' so to say – despite the fact that, physically, she spent many weeks just learning to walk again, first with the aid of a zimmer, and ever afterwards, a stout stick. And there was also the crucial matter of learning to speak again - amazing for us to witness. Maybe - but for Mary? Not really. In those twenty years of serious disability that followed, she never, ever, showed even a wisp of self-concern. Instead, she continued to be as much concerned as she had ever been for my well-being, and, indeed, the well-being of any and all of those whose lives touched hers, during the remaining twenty years of her life.

And it was this quality of spirit in her that made it possible for me to continue with, and to develop, my new career at the NPL. As already acknowledged, the other side of that coin was the staunch support of my new colleagues, both professionally, and as personal friends. In their own way they had qualities almost as special as Mary's, so that, even forty years on, as I look back on those early years in my new job, there is a certain aura about them, of - what shall I say? – responsible carefreeness – freedom from the petty cares of ordinary life, in the face of their responsibilities as the custodians of national and international standards of the widest possible nature and importance. One might almost draw a parallel between those who, in the monastic life, have taken upon themselves the custodianship of moral and spiritual standards, also finding themselves freed of the pettiness of ordinary life. Fanciful? Maybe – maybe not.

Be that as it may, the fact is that Mary and I seemed to pick ourselves up within a matter of months, and go on our way again – no – not exactly rejoicing, but each of us, in our different ways, realizing that our lives had moved into an entirely new state: what was to be 'normal' for us in the future was never to be the normality of the past. For one thing, physically, Mary would never be 'running the household' again, as she had done, so devotedly, and so efficiently, for a quarter of a century. But she was to continue to fulfil her role as 'hub' of the family even more influentially than before. What had happened to her had, of course, *physically* 'disabled' her. But

Mary, being Mary, saw to it that, in other ways, it had *en-abled* her – to share with all those around her something of the tranquillity of spirit that was to characterize the remaining twenty years of her life: to share that 'wink of Heaven' she had had.

On the practical level, for months (or was it even a year, or more?), she would be unable to undertake even the simplest of household chores, and, as in many another context, it was our friend Tom who provided the answer: he suggested that one of his parishioners, Julie (I can't remember her surname), whom we already knew as a co-member of Tom's congregation, would probably be more than willing and able to help us out.

And so it was that I was soon able to leave home as usual, and at the usual time, to jog down to the station again, on my way to meet the challenges which, daily, awaited me at the NPL.

I have just seen my move to The National Physical Laboratory in another light: it was as though I had pressed a button like the one on my computer, depicted as a file that is just being opened, attached to the flap of which is a broad, upward-pointing arrow. If you rest the cursor on it, it says 'Up One Level' And that's just what it felt like! I don't mean salary-wise, or even status-wise – I mean function-wise. You see – as a medical physicist responsible for the control of radiation dosage given to radiotherapy patients, one employs a so-called 'secondary standard dosemeter' for the measurements, and in Great Britain that secondary standard needs to be calibrated regularly by the National Physical Laboratory. So, for me, it was indeed 'up one level' in the dosimetic hierarchy - from being a user of secondary standards, to being responsible for their calibration against the primary standards. And, intriguingly enough, I was to be in charge of the basic standard of X-ray dose in an earlier version of which I had found fault some ten years before... Full circle, one might say.

During my first couple of years in post at the NPL, I continued to be involved - on a consultative basis - with my former colleagues at The London Hospital, to enable projects still ongoing when I left to be brought to fruition. As a result, a number of papers appeared in the journals over that period which seemed to suggest that they were the outcome of collaboration between my old Department, and my new one at NPL. In reality, though, it was just a matter of 'tying up the loose ends'. The titles of the papers concerned, included innovative work on solid-state circuitry – still, then, in its relative infancy – and its application in various radiation measuring devices and monitors.

With this rear-guard action completed, it was the most natural thing in the world to turn one's attention to those secondary standards, for the regular calibration of which my new Department was responsible: why should they need to be re-calibrated, over and over again – in fact, every year or so? The answer was not difficult to come by. They utilized one of those so-called 'thimble' ionisation chambers, which the hospital physicist would place in the radiation beam used for treatments, to ascertain the number of units of dose ('Röntgens') delivered by the beam in a measured time: hence 'dose rate'.

Now - the thimble-shaped 'cap' of the chamber was made of a so-called 'plastic' material (was it bakelite?), coated on the inside with graphite to make it conducting. The trouble was that the plastic material was, geometrically, not very stable, with the result that the volume of air inside it could change in volume by a significant amount over the comparatively short period of a year or so – hence the need for such frequent re-calibrations at the NPL.

I asked myself, 'Was there a material much more stable geometrically than the plastic, which could replace the latter?' The answer was staring me in the face: graphite itself!

Instead of being moulded, like the plastic caps (*moulding* not being the best starting point for long-term geometrical stability) the caps could be turned out of solid graphite on a lathe, guaranteeing an extremely stable shape. So, we designed a completely new secondary standard dosemeter employing a new graphite-capped thimble chamber, to replace the existing secondary standards in use, nationwide. The new secondary standards were production-engineered by Nuclear Enterprises, Ltd, and some thirty of them manufactured for use in designated centres across the country.

But it didn't stop there. One such instrument was taken on a world tour to a total of 22 national centres, to carry out on-the-spot check calibrations of the local standards in use. the visits also including visits to seven national standardizing laboratories, the instrument thus playing a part in ensuring world-wide uniformity in radiation dose.

Almost two decades after I had retired, I received a copy of a long-term study of the stability of the new secondary standards: *they had shown no significant change over the whole of that period!* – though, for legal reasons, they had continued to be calibrated regularly. Surely, a gratifying outcome to that hunch about the use of a solid graphite cap.

Everything I have described so far, concerning both primary and secondary standards, has related to the use of X-rays at the levels of dose and dose-rate employed for treatment purposes. However, where radiation is in use at such levels, there is inevitably some escape and scattering of radiation into the work spaces occupied by those involved in administering the radiation treatments. This creates a potential occupational hazard in the long-term, which needs to be monitored by specially sensitive dosemeters capable of measuring the low levels of radiation involved. This, in turn, involves the use of ionisation chambers very much larger than the 'thimble' chambers we have been describing, in order to generate measurable currents. 'No problem,' you might think, but such work-space monitors need to employ not only very large chambers, but ones with very thin walls. NOT so easy!

Then came another one of those hunches, which are as

necessary in science as in any other branch of human activity. During my 22 years at The London Hospital I had naturally enough become familiar with much of the equipment in routine use by my medical colleagues, though having no direct involvement with it myself. One such piece of equipment was the so-called 're-breathing bag' which was part of the anaesthetist's stock in trade. Made of thin latex rubber, it was impregnated with graphite to make it conducting, thus avoiding the accumulation of static electricity and the accompanying risk of anaesthetic gases exploding. The re-breathing bags were ellipsoidal in shape, but there was no problem in getting spherical ones manufactured, of several litres' capacity. The bag was then mounted on the end of a stem and inflated to a pressure slightly in excess of atmospheric, encasing an electrode which collected the (electrically-charged) ions produced in the enclosed gas, which could then be measured, and the level of stray radiation present deduced. It was a simple, and effective solution to a potentially difficult problem, and, as far as I know, is still in regular use (Plate 11).

The clinical secondary standard X-ray dosemeter, together with the balloon chamber for the measurement of low-level stray radiation, and the Linear Radium Source Dose Calculator, all comprised major advances in the field of dosimetry, for which, in 1973, the Worshipful Company of Scientific Instrument Makers presented me with their (annual) Achievement Award, the official citation - that it was 'for [my] life-long work in the field of radiation dosimetry, culminating in the NPL Secondary Standard Therapy Level Dosemeter.'

Apart from a Certificate to this effect, the Award included a beautiful trophy in the form of a bronze replica – mounted on a black onyx base - of a head of Minerva, found near Rome, *circa* AD 150 (Plate 12), a reproduction of which, incidentally, crowns the coat of arms of the Worshipful Company.

I found myself asking, 'Why Minerva?' According to the Encyclopedia Mythica, she was - among other things - the Roman goddess of wisdom, medicine, the arts, dyeing, science,

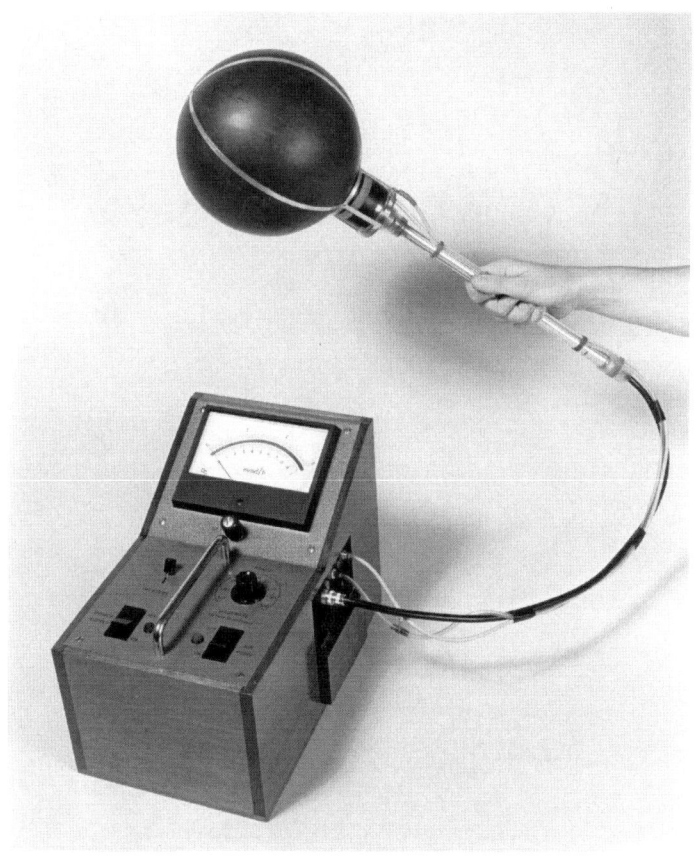

Plate 11 Inflated Balloon Chamber and Associated Electronics

Plate 12 The Worshipful Company of Scientific Instrument Makers Achievement Award 1973

and trade, as well as being the inventor of numbers, and musical instruments. I suppose that - the list being as comprehensive as that - it would not be unreasonable to suppose that had there been scientific instruments in Roman times, they would be likely to have come under her aegis!

Chapter 17

Enough science and technology for the time being! Let us return to the Sixties - and had I been titling these chapters, this one would certainly have earned the title, 'All Change for the Family'.

From the perspective of fifty years on, those ten years - from 1960 to 1970 - look very brief, and, in one way or another, and, from the same perspective, they certainly appear to have brought a whole series of life-changing events to each and every member of the family.

However, I need to remind myself that they were not, as it seems now, merely ten 'short' years, but significant fractions of their lives for our family: add ten years to the life of a late-teenager, and you have added fifty percent to his or her life! And in the case of each of our children, those ten years included A-levels, choice of a career, graduating (and, in John's case a Ph D, in Roger's. a six-year Architecture course), becoming engaged (yes, one did so in those days...), married (that also still a norm), and starting their careers. Phew! Yes! *Phew!*

Let's not forget that I, too, had changed my career in the middle of this period. And my poor, dear Mary had suffered the stroke which changed her life for ever. It's time to fill in the details.

John led the way – finishing his schooling in 1960 with five A-levels to his credit. As already noted, these were Physics, Chemistry, Zoology, Botany, and Mathematics, the latter very much the odd one out, but what a basis the other four, for the understanding of the nature of the processes involved in what constitutes 'life'! Mathematics he took no further, and it has remained something of a closed book to him ever since, but in the Autumn of 1960 he took up his place at Bristol University to study Honours Chemistry, with Physiology as his subsidiary subject. He duly graduated in 1963, and went on to Manchester University to take a PhD in 1966, the title of his thesis, 'Polyfluoroaromatic Azo-compounds, and some related heterocycles' - to which my response, as a physicist and (something of a) mathematician too, has to be 'Phew!' again – the other chap's jargon always being much more impressive than one's own.

Continuing John's Sixties' saga, in 1965 he met up with a fellow student, Catherine, whom he married in 1967. His chosen career was drug research, and he obtained a post at Pfizer's, in Sandwich, Kent – one of the largest drug companies in the world. This he took up in 1968, after he had completed two post-doctorate years, and Catherine had finished her PhD. So it was that the newly married couple made their first home in Canterbury. As it transpired, John was to spend the whole of his working life at Pfizer's – which must be regarded as quite unusual in these days, when modern transport, and, in particular the motor car, has made the work force so much more mobile. It must be remembered, however, that the highly specialized nature of his work considerably narrowed his choices.

Our respective disciplines were. of course, poles apart, which meant that neither of us could appreciate in any detail what the other was up to: but there, as they say, variety is the spice of life. And that being so, there was to be plenty of spice in the Kemp family: John a chemist, Roger an architect, Rosemary a musician, Mary a linguist and kitted out for social service work, and your humble servant a physicist-cum-mathematician.

But I anticipate. Roger, two years younger than John, duly left school in 1962, to study architecture at Liverpool University, one of the leading Architectural Schools in the country. A first degree in architecture takes as long as a first degree in medicine, which is understandable, since designing and erecting an insecure building is as dangerous to life (maybe more so) than an incompetent doctor...

Roger met Wendy, a language student , very soon after going up to Liverpool, and they became engaged in 1965, and married in 1966. Wendy was teaching by then, and Roger had two more years of study before graduating, so – unlike John and Catherine – their first two years of married life were spent in a small flat, on site, so to say, in Liverpool. After graduating in 1968, Roger obtained his first post with an architectural firm based in Bath, and they set up home in Coleford, a typical Somerset village a few miles out of Bath.

Five years younger than Roger, Rosemary was a late starter in the Sixties' Stakes, but one might be forgiven for thinking that she was determined to do her best to catch up with her brothers, even though it was 1967 before she left school. By that time she had become very interested in music therapy, so, with this in mind, she enrolled as a student of La Saint Union College of Education at Southampton. It was during this period that, following a complex chain of events which need not concern us here, a young German teacher – one Wolfgang Weiss – became a friend of the family, and friendship quickly turned to romance for Rosemary, and she and Wolfgang were married in Germany in April 1970 – Rosemary returning to England to see out the last term of her Education course.

It need hardly be said that her marriage (at 21) resulted in some basic changes in her musical education, the location of which of necessity switched to Germany. There she continued her studies with Siegfried Raabe, of Hanover, principal flautist of the Orchestra of North German Radio, and Willi Leenen, who was the organist at Minden Cathedral. In fact, even as I write, it has come to mind that I still have a tape recording of her, in the Cathedral, and accompanied by the organist, playing 'Dance of the Blessed Spirits', by Glück, and a movement from a Sonatina by Lennox Berkeley. Well do I remember my excitement at being able to record these, many hundreds of miles away from the actual performance: you see, the Service was broadcast! And the date? – 1972. She subsequently came back to England in 1974 to take the 'Licentiate of the Guildhall School of Music' examination, and went on to teach flute, part-time, at the Minden School of Music, at that time just starting up. And in 1980, for reasons which we need not deal with here, she moved south to live near Stuttgart, and became a full-time flute teacher at the Lower Rems Valley School of Music, where she has remained ever since.

I have strayed far beyond the Sixties, but in view of the complexities which came upon Rosemary's music career following her move to Germany, it seemed appropriate to do so. Even so, there are plainly some gaps in all this which we may have the opportunity to fill in later.

Was that it then? I mean – as far as our family and the Sixties were concerned. Strictly speaking, yes – it was But how were we to know that within very few months after the decade had ended, we were to have – albeit not in the literal sense – what, nevertheless, came to feel very much like an addition to our family!

It was a process rather than an event, which, all unbeknown to us, had begun on August 10th, 1969, when a young Australian, Bruce Hedland-Thomas, and his wife Lyndy sailed from Freemantle, their destination Tilbury, where they docked on September 6th. They planned to spend two years in England, during the first year of which Bruce would be attending an MSc course at Surrey University, on 'Radiation Studies'. Yes – it could be said that their first link with us had been forged several years previously, when Bruce had taken up Medical Physics; and that link was brought almost to our doorstep by his having chosen Surrey University, situated on the same hill as Guildford Cathedral, and little more than half a mile as the crow flies from our home in Thorn Bank. And the link was subsequently brought to within a few hundred yards by their taking up lodgings with a certain Mrs Patient, a middle-aged widow who also lived in Onslow Village. It was in fact she who had introduced us to Wolfgang, Rosemary's husband-to-be. As Bruce put it, 'When Mrs P discovered that I was a hospital physicist, she started to insist that she would introduce me to another physicist who lived in Guildford, adding that she 'had boarded a German student who had married his daughter' As already implied, the link was finally forged some months later.

One requirement of Bruce's MSc course was that students from foreign parts had to give a presentation about the departments from which they came, at a specially arranged seminar. The seminar took place in the Spring of 1970 at Surrey University, and it was there (having attended the seminar) that I invited him and Lyndy round to afternoon tea. There, of course, they met Mary for the first time, still recovering from the stoke that she had suffered nearly a year before. The significance of that occasion was not lost on Bruce either: as he put it, 'It's not for me to speculate, but something happened for that teatime not to be an isolated event. Lloyd and Mary became *in loco parentis* to us. Our lives became intertwined up until the time we left England in 1976 - and beyond, to the present day.'

A little later that year Bruce and Lyndy moved into a married students' flat at the University – just a little further away, but still barely a mile from us, which meant that we were able to continue to see them relatively frequently. In fact, with Roger and Wendy in the SW, and John and Catherine in the SE, and Rosemary in Germany, we saw a great deal more of them at that time than we did of our own children, despite the fact that our first grandchild, Anna, had been born to Roger and Wendy on February 5th, 1970, and our second grandchild, Vanessa, to John and Catherine, on the 21st of July, 1971. Little wonder that when Bruce and Lyndy's first child (Genevra) was born on the 3rd of August 1971, we felt almost as though a third grandchild had arrived hard on the heels of Anna and Vanessa! – a feeling shared by Bruce. Apparently. within an hour of the birth, he met Mary and me, out for a little walk, so, as he put it in an *aide memoire*, 'They were the first to know - *appropriate for adopted parents.*' (Italics mine).

Despite the major distraction of becoming a father, Bruce duly obtained his MSc, and the Award Ceremony took place on July 15th 1972. There was one regrettable but inevitable consequence – that they had to move out of their University flat, to a basement flat in what Bruce called 'a grand house on Ham Common, in Richmond': inevitably, alas, it meant that we should see less of them.

Over this period, Mary had continued to make steady, and in many ways truly remarkable progress. Despite the fact that within a year of the stroke she had undergone an operation for gallstones, we flew to Guernsey for a holiday in the summer of 1969, and a few weeks later to Germany, to take part in the engagement celebrations of Rosemary and Wolfgang. And, as already noted, with the help of our good friend Julie, and the encouragement and stalwart support of Tom, our vicar, life, in many ways, had returned to something approaching normal again, as we entered the 70s. It meant that my mind was free enough to contemplate breaking new ground in my work at the NPL - and it was with the encouragement embodied in a letter I had received in June 1969 from the Administration at the Laboratory (Plate 13).

[Regarding this promotion, only days ago, rummaging around among old documents and papers, I stumbled on a copy of the text (\Appendix 4) of the submission which had been made on my behalf in support of the special promotion, and reading it through, 40 years on (I am writing this in 2009), was a strange and, indeed, novel experience for me. I must, of course, have read it in 1969, when it 'was all happening', so to say, and though it must have been gratifying to do so, it would not have had the effect on me then, that it has just had, after 40 years.

In 1969, my career still had almost 10 years to run. It had been, and still was, an ongoing series of contiguous, but seemingly mostly unrelated projects: the work on the radium source dose calculators, followed by an 'ionization current comparator' which, among other things, made an 'automatic dose plotter' possible; then the work on the international standards, followed by the move to the NPL. and the work on the secondary standards, and so on, and so on. Not for one moment did I glimpse 'the big picture' which it was all 'adding up to'. And if I hadn't glimpsed it then, how much less likely was I to have glimpsed it now, when I have long since moved on to other things, and, so to say, 'lost the flavour of it', years ago?

From: G. Wheeler, CB, Principal Establishment Officer

MINISTRY OF TECHNOLOGY
Shell Mex House, Strand, LONDON W.C.2
Telephone: 01-836 1207

HC/48/016 Q

23rd June, 1969

Dear Dr Kemp

You will be pleased to know that approval has been given to your promotion as an individual research scientist to Senior Principal Scientific Officer. This promotion takes effect on 1st July, 1969; details of your revised salary will be sent to you in a separate letter.

I should like to add my personal congratulations.

Yours Sincerely

Geoffrey Wheeler

Dr. L.A.W.E. Kemp
Through D/NPL

Plate 13 A Shot in the Arm!

So what *was* this 'novel experience' that I have just undergone? *Please* don't misunderstand me: I haven't suddenly acquired a swollen head - but what I *have* acquired - and for the very first time - is some idea of what this whole series of projects must have looked like *to my professional colleagues.* I repeat: my head is the same size as it ever was, but for the first time *ever* I have seen my career in medical physics 'in the round'. With no 'master-plan' in mind, I had *in point of fact* - over a period of nearly 30 years – covered the *whole gamut* of clinical dosimetry, from arrays of radium needles and radon seeds used as implants, to X-ray and gamma-ray beams, ranging from those used to treat superficial skin lesions to those used to treat deep-seated tumours. With the completion of this phase, I had moved on to the primary (international) standards of dose, only to find long-standing errors in them, inevitably, then, becoming involved in their elucidation, and elimination. Ensconced in my post at the NPL, there remained the so-called 'secondary standards', linking the primary standards to day-to-day radiotherapy practice, with new designs providing a highly desirable long-term stability in their performance.

This is the 'big picture' then, which, as I say in all truth I have only *now* come to see 'whole'. And, grotesquely, if credit is due, it must ultimately be given to my mother's cancerophobia – yes, I mean that – for I know in my deepest self that *that,* and that alone, could have given me the necessary motivation and impetus, over the years of effort necessary to see such a mission through. What a metamorphosis! Strangely, I don't remember ever discussing my career with my mother, even though she lived to see the bulk of it unfold (she died in 1963). But there – who knows? - it could well have been a 'no-go' area for her. What I *do* know is that without her cancerophobia, *plus* a world war, that 'big picture', such as it is, would never have been painted, and I would not have been sitting here at my computer, looking back over the ninety-five years of my life, writing such a book as this. A notable case of 'good out of evil'? Could be. God's alchemy? I wonder.]

To resume then: what *was* that new ground that I found myself contemplating in the middle of 1969?

The so-called 'free air' primary standards providing a calibration of the secondary standards employed in hospitals in terms of a unit known as the 'Röntgen' (named after the German physicist who discovered X-rays just before the close of the 19th century). Basically, this involved measuring the electric charge set free by the radiation in interacting with a cubic centimetre of air at standard pressure and temperature. In many ways a more basic approach to radiation dose measurement is to express the dose in terms of the amount of energy deposited by the radiation per unit mass of any particular material under consideration. The corresponding unit of 'absorbed dose' is termed the Gray (after Hal Gray, an English physicist working in this field in the 20th century). 1 Gray corresponds to the deposition of 1 Joule of energy per kilogramme of the particular material concerned, the important point being that this will be the same, whatever the type of radiation (X rays, electrons, alpha particles, neutrons, etc., etc.). Standardization in terms of 'Grays' (instead of 'Röntgens'), expressed in basic terms, involves the accurate measurement of a very small rise in temperature of a known amount of a standard material, the corresponding instrumentation for which is known as a 'micro-calorimeter.

In a nutshell then, I moved from the measurement of the very small electric charges released in air by the radiation, to the measurement of the extremely small temperature rises produced in a solid material such as graphite. And to my delight, the preliminary work would be theoretical, and highly mathematical. Indeed, all that hard work I had done, working through L A Pipes' book as I commuted back and forth on the Southern Railway was to come into its own at last. And not only that, but the analysis of the transient behaviour of the particular kind of micro-calorimeter I had in mind would find me up to my neck in those wonderfully elegant and indeed incredibly beautiful 'Laplace Transforms' which had so captivated me, as I journeyed to and fro.

You think I've gone a bit 'high' on Laplace Transforms, and the like? Then take a look at some of the equations to which they led me, and the family of curves that emerged from them. Is there not real elegance and, yes, beauty to be discerned in those curves? How come such beauty from such ungainliness?!

Eqn.(43): $\theta_a = \dfrac{V_I}{A}\left[t + \left(\tau - \dfrac{1}{AB}\right) + \dfrac{w}{AB}\left\{1 + \dfrac{(T-AB)}{(S-T)}e^{-St} - \dfrac{(S-AB)}{(S-T)}e^{-Tt}\right\}\right]$

Eqn.(45): $\theta_a = \dfrac{V_I}{A}\left[t + \left(\tau - \dfrac{1}{AB}\right) + \dfrac{w}{AB}\left\{1 - e^{-\frac{t}{2\tau}}\left|\dfrac{(1-2AB\tau)}{\sqrt{4AB\tau-1}}\cdot\sin\dfrac{\sqrt{4AB\tau-1}}{2\tau} + \cos\dfrac{\sqrt{4AB\tau-1}}{2\tau}\cdot t\right|\right\}\right]$

Eqn.(46): $\theta_a = \dfrac{V_I}{A}\left[t + \left(\tau - \dfrac{1}{AB}\right) + \dfrac{w}{AB}\left\{\left(1 - e^{-\frac{t}{2\tau}}\right) - \left(\dfrac{1-2AB\tau}{2\tau}\right)te^{-\frac{t}{2\tau}}\right\}\right]$

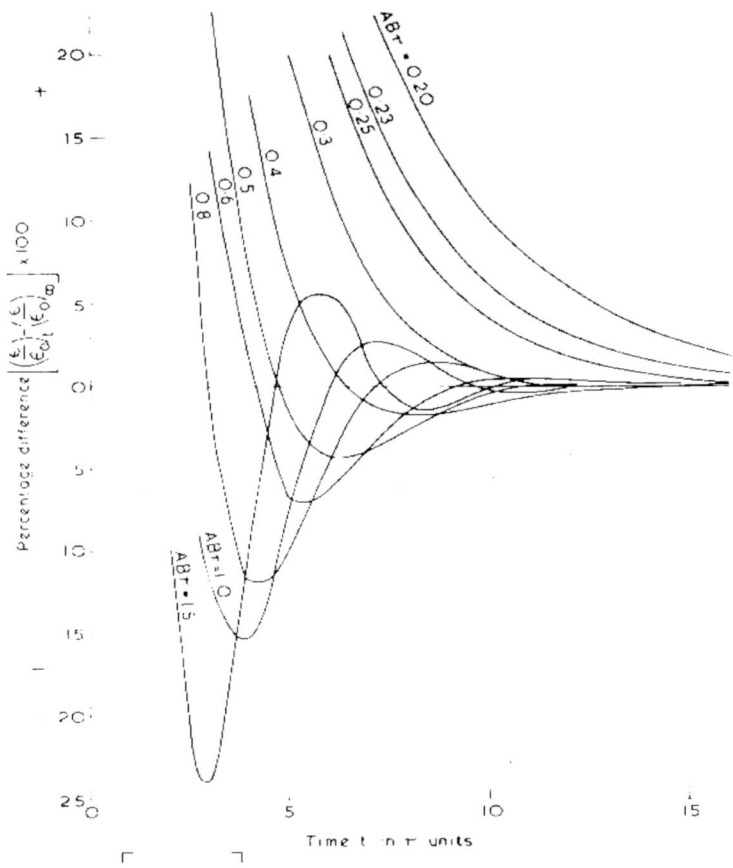

OK. You're right, I suppose - beauty *is* in the eye of the beholder, and especially when it comes to mathematics...

The above equations, and the curves to which they gave birth, were taken from a Departmental Report written in April 1976, when I had reached the peak of my career at the NPL, and within a year of two of retirement. The rest, one might have thought, would have been 'downhill all the way'. But alas, not so.

I had a close friend, some forty years ago, who, memorably, once said to me, 'You know, Lloyd, life is inclined to give you a pat on the back one day, and a smack on the bottom the next.' And, that April, when, yes, I was riding high, little did I know that life was preparing to deliver not so much a smack on the bottom, but a terrible body blow, as I made my way towards retirement. The irony of it was that, within but a few weeks of the blow falling, I was to learn that a totally unexpected slap on the shoulders was on its way.

It was on May 2nd that I had picked up the post off the front-door mat, to find that, amongst a number of nondescript envelopes, was one distinctly official-looking, which I immediately took into the bedroom to share with Mary.

Looking back on the mid-nineteen seventies, I have to say that they had seemed almost halcyon days, having flown to Germany again in 1976, to visit Rosemary almost within days of the birth of her first son, Jeremy; and on that morning of May 2nd we were looking forward to motoring to the Lake District in August. Even so, nothing could have prepared us for what was in that envelope! It was an official notice on behalf of the Prime Minister, in effect asking if I would accept an OBE, in recognition of my 33 years' work in the field of radiation dosimetry. It was a complete 'bolt from the blue', and I can still see, in my mind's eye, Mary sitting up in bed, and myself standing beside her, each with a hand on the letter. The Award ceremony was to be later in the year — but what a prospect, with a holiday in the

Lake District to precede it!

But, all too sad to say, the delivery of that body blow was to follow the publication of the Queen's Silver Jubilee Honours List within a matter of days, and we would never have that holiday in the Lake District; and Mary would never be able to look forward to accompanying me to Buckingham Palace.

It Happened in the middle of the night of Tuesday, June 21st. I had thought at first that we were making a routine middle-of-the-night visit to the loo, until I heard Mary saying what a fine young man our son Roger was, and how grateful she was that he was now in Heaven. All the horror of the realization that she was having another stroke came crashing in on me. Having got her back to bed, she soon sank into a coma, and the next morning our doctor, on a second visit, discussed with me the pros and cons of admitting her to hospital. He concluded what he had to say by volunteering his opinion that if *he* were in Mary's position, he knew what he would prefer. What he was trying to tell me, as gently as he could, was that, if he were dying, he would want to die in his own bed.

However, only two days later, he rang me to say that he had been discussing Mary's case with a specialist, and they thought they *might* be able to help her: was I willing for her to be admitted? 'Of course,' I said, unaware that I had just spoken a mere two words which could mean death for Mary, rather then new life. How fortunate it is that, sometimes, we are not aware of the full weight of the responsibility resting on our shoulders: we might seize up altogether – failing to act at all – and, by default, usurping altogether, those very responsibilities. As it was, I was allowed to visit Mary the same evening, and there she was, sitting up in bed almost as though nothing at all had happened. And they say the age of miracles is over...

My 'gong' (it's the 'in' word, so I'll use it!) was duly pinned on me one day in November of that year by Her Majesty, witnessed by John, Roger, and Rosemary - who had come over from Germany specially for the ceremony (Plate 14). And, within

Plate 14 En Route for the Palace (seen in the background)

a year, I had retired.

At the time of my retirement, I received a formal letter from the Permanent Secretary at the Department of Industry, dated 28th March 1978, marking the occasion - and looking at this, and a copy of my reply, it seems that we were on first-name terms...

And scrutinizing my reply, it would appear that his letter had produced a similar reaction to that which I had experienced in response to the letter I had received at the time of my Special Merit promotion - I just didn't recognize the picture of myself that he had painted:

'Frankly, I didn't recognize myself in your description of me, and of what I have tried to do since I came to the Lab 12 years ago! And the reason for that is. I think, quite simply that whatever has been done during those 12 years has been very definitely the fruits of the scientific 'climate' which first you yourself, and latterly Alan [Jennings], created for me in which to work, and most assuredly also the solid teamwork of those whom I have had the privilege to work with, over that time. I cannot begin to say what it meant to me to leave behind the 'cut and thrust' of hospital life (an unfortunate metaphor!) to spend the last phase of my career in such a congenial atmosphere as I found waiting for me at the NPL, and with a bunch of colleagues unequalled anywhere for the loyalty and deep personal friendship which they offered me from the start.'

So ended my 34 years of active service in the front line of the war against cancer – a very personal war, as I have explained. I hope no-one would begrudge my claiming to have 'fought the good fight', which I certainly would not have been able to claim, had I stayed in a so-called 'reserved occupation', way back in 1939.

Chapter 18

And now, at the beginning of this chapter - having surveyed the main road again - I find myself about to explore yet another lane, this time (and somewhat enigmatically) it might be called 'Learning to Drive a Car - and the Morse Code'.

We are back at the very beginning of the 70s, with Mary, two years on from the stroke, slowly, but steadily, gaining in both strength and mobility. Despite that, many things which, in normal circumstances, we would have undertaken with hardly a second thought were still barely possible for us to contemplate. And strangely indeed, at that time, it was actually easier for us to go to Heathrow, and take a plane to Germany, than to embark on a holiday in England, which involved negotiating our way onto railway platforms via staircases and subways, and, subsequently, maybe having to change trains to complete our journey. In fact, one could say that our long-range mobility at that time had far outstripped its more parochial counterpart.

Apart from a spell just after the war, we had always lived in the country, and each member of the family being well equipped with two good feet and legs and all of us enjoying country walks as we did, we had never felt the urge for a car. But, in the circumstances in which we found ourselves after Mary's stroke, learning to drive so that a whole new dimension could be added to Mary's life - well - that was quite another matter!

So it was that (unlike the youngster of today, who almost takes it for granted that immediately after his 17th birthday he can expect to have driving lessons) I found myself taking my first driving lessons at the tender age of 57. I did actually enjoy learning to drive late in life and, in my advanced years now, I can still remember the thrill, comparable with that of Mr Toad himself, which I felt when, for the first time, and with no-one else in the car, I drove through the traffic in the centre of town.

It does make me wonder how many of today's youngsters will enjoy such a recollection in their old age...

I did very much enjoy, too, being Mary's chauffeur for all those years that followed – nearly twenty of them. She would sometimes say, as she sat beside me in the car, 'You're a good driver, you know,' but in the nature of things that certainly couldn't be taken as any sort of a realistic assessment of my driving skills! It was, I am sure, merely her way of saying that she was happy to be out for a ride, and perhaps, as well, that she felt safe with me. Come to think of it, in those terms it probably did represent a judgement on my driving—it may even have been a sort of compliment after all - based on pragmatism!

Perhaps another advantage of learning to drive late in life was that I was aware, from the very beginning, that a car is a ton of lethal weapon, and over the twenty years that I was driving, I never lost that right and proper awareness. However, when Mary had died, and the imperative for being on the road in a car was removed, I suddenly became aware of the jungle that modern traffic comprises, and the jungle warfare that so often goes on there, and that 'proper' awareness quickly became an untoward one, so that I began to feel anxious and apprehensive all the time I was on the road. I gave up driving then (in 1991, when I was 77): the motor car had served its vital purpose in our lives.

I had never had more than a slight brush with any other vehicle, and by giving up driving when I did I am able to look back on that period of our lives and feel that the blessing that the car had been to us had never become a blight to any one else. I sold B 164 WOU privately, and as its new owner climbed in, for the first time the car seemed to acquire a personality, and I felt I was losing a friend. I looked at the seemingly empty passenger seat in the front, and in my mind's eye I could see, so plainly, Mary sitting there.

But what's this about the Morse Code? Ah - that's *quite* another story... At the risk of appearing to 'fly off at a tangent', let me begin with the reminder that when Bruce and Lyndy sailed for England in 1969 their intention had been for a two-year stay, which, in the event, was extended year on year into the mid-70s, by which time their second child Rebecca ('Becky') had been born, and Mary and I had become well-established as effectively a part of this young and happy nuclear family. In general, I don't allow such sayings as 'All good things come to an end' to influence unduly my views on life, but I confess that when Bruce announced that they would be returning to Australia permanently, in April, 1976 I gave voice to some such thought, adding (to soften the blow, I suppose) that 'there was nothing much that we could do about it.'

To my great surprise then, and looking me straight in the eye, Bruce said, with sudden enthusiasm, 'There *is,* you know!' For a moment I couldn't believe he really meant it, but another glance at him assured me that his enthusiasm was genuine. 'Come on then. Out with it,' I said, infected now with his enthusiasm. 'What about both of us becoming radio hams?' he responded. I wasn't familiar with the radio ham world, but I knew enough to realize that there was more to becoming a radio ham than filling up the necessary application form, and paying some sort of registration fee to obtain a licence. So what lay behind the suggestion? I thought I had an inkling. Remember – it was 1976, not 2009 (the time of writing), when I am on a 'TalkTalk' telephone tariff which, for the line rental plus £10 a month, enables me to talk to *any*one, *almost* anywhere, at *any* time, and for as *long* as I like. (I talk to Bruce for up to an hour sometimes, and on average several times a week.) This time 'Google' has failed to come up with the goods, so I am having to guesstimate that in 1976 it would probably have cost around £1 a minute for calls to Perth, which would be the equivalent of several pounds a minute today. As radio hams, however, we would be able to talk to each other – yes, *for as long as we liked,* though not quite for free. There would be an annual licence fee, and the initial capital outlay of hundreds of pounds for the equipment (equivalent to a couple of thousand or so now). And it wouldn't be *whenever* we liked, either. There was the little matter of an 11-year sunspot cycle to reckon with (long-distance transmission being at its best at sunspot maxima, and usually difficult when they are at a minimum - all to do with those ionised layers in the upper atmosphere). Added to that, and quite apart from the influence of sunspots, there was the somewhat fickle behaviour of the ionised layers even at the best of times, the layers being involved in a multiple reflective process which enabled radio waves to traverse the curved path round the Earth. Even so, however, there was no doubt that Bruce's proposal was an inspired one, which immediately put an entirely different complexion on their return to Australia: it offered good prospects that we could remain in good touch, despite the 9000 miles that would separate us.

There was a full year or more before their departure in April 1976, and we both set about acquiring the necessary City and Guilds certification which would enable us to acquire our licences. This was no sinecure, and involved passing an exam on radio transmitter and receiver theory and design, as well as the mechanisms involved in the propagation of radio waves. As if that weren't enough, there was also a practical test involving transmitting and receiving messages in the Morse Code at a minimum of 12 words a minute! You might well ask *why*, when we should be talking plain English to each other. It's true that, half a century earlier, Morse code would have been essential if one wished to make long-distance contacts. But with the steady improvement in receivers and transmitters, and especially the huge increase in radio-frequency power provided by the latter, the use of Morse had become a sentimental matter confined almost entirely to the 'old-timers', with (dare I say it?) a little bit of 'showing-off' thrown in for good measure.

So why *were* the likes of Bruce and me subjected to such a torment? It was quite simple really: *it was to sort out those who were serious about getting the licence, form those for whom it was but a passing whim, so to say.* How so? Well, you see, becoming proficient in the Morse Code – even up to the modest speed of 12 words a minute – was a really hard slog, and all but the most determined fell by the roadside. (I have always maintained that it was pity for my grey hairs that tipped the scales for me!)

Bruce studied on the train, as he travelled back qnd forth to his job at Guy's Hospital, and I used to break my journey from Surbiton to Guildford at Brooklands, where there was a night school that catered for budding Radio Hams. In fact, by the time Bruce and Lyndy finally departed, we were both qualified, and all that remained was for us to acquire the necessary equipment, and choose, and erect, an aerial system capable of long-distance working, and we would be in business! When I say '*all* that remained', it was indeed still a big 'all', but it was cut down to size by the prospect that in the not-too-distant future now, we would be talking to one another from our back gardens, so to say, despite the thousands of miles that would separate us. *What* a prospect!

Bruce tells me that that historic first contact took place on the 3rd of January, 1977, it having been agreed, it seems, that I should initiate the call, a procedure which would have run something like this:

'This is Golf Four Delta X-ray Lima calling Victor Kilo Six Oscar Oscar,' repeated two or three times, and then, 'This is Golf Four Delta X-ray Lema calling Victor Kilo Six Oscar Oscar *and listening*.'

And THEN came the magic moment: 'This is Victor Kilo Six Oscar Oscar returning your call.' I really did feel like Marconi must have done on that historic day in December, 1901, when, ensconsed (and - yes! – with an assistant named Kemp!) in a receiving station on the coast of Newfoundland, he heard the three 'dots' of the letter S in the Morse Code, transmitted by his assistants in Poldhu, in Cornwall, 2200 miles away. Scientist that I am, there is a sense in which I still retain the word 'miracle' in my vocabulary to convey the depth of my feelings on such an occasion, when something I had initiated in my back garden had been detected by Bruce 9000 miles away in *his* back garden, *just half a second later.* Please, Lord, let me never lose that sense of wonder! - it just isn't sufficient to know *how* it all comes about, it's the fact that the material making up this world of ours has been endowed with such wondrous properties that such things are, indeed, possible at all...

For the record, Plate 15 is a photograph of me at the console of G4DXL: it occupied nearly half our spare bedroom, my comment to guests being, 'I'm afraid you're going to have to sleep with that.' But just look at all those knobs and dials! - it was 'home from home' for me!

And of course, it didn't stop there – I mean, simply as something Bruce and I had organized for ourselves as a convenient and cheap means of communicating with one another: we had in fact each acquired a brand-new hobby, of which it might be said that 'the world was our oyster' – an 'oyster', withal, which could be entered without stepping outside the door. Thus, instead of calling a specific individual (using his call sign), one simply said, 'Calling CQ. Calling CQ,' - one of the numerous time-saving abbreviations used by devotees to the Morse Code. And if one was looking for a far-off contact, one called, 'Calling CQ DX,' – DX being the abbreviation for 'long-distance'. And so on, and so on.

And it was a bit like fishing: one cast the line, so to say, but one didn't know who would take the bait! And THAT was what made it such fun, and really exciting. Well do I remember a number of such calls, decades after, simply because of the surprise that they were, at the time. There was for example, the radio operator on a cargo ship in the middle of the South Atlantic. He was off-duty, but had put on his other hat as a radio *ham*. One might have thought that, off-duty, he would have been a little more enterprising than to choose 'more of the same'; but that's just the point! – it was *not* more of the same. Radio hams go out of their way to avoid the formalities of the

Plate 15 'Calling CQ'

professional. The world of the Radio Ham is much more like a brotherhood, in which you are on first-name terms from the word 'go', and will end even your very first contact (with another ham) with '73' - going all the way back to 1859, when the Western Union telegraph code was standardized: in that code '73' signified 'Kind Regards'. And if you are well acquainted with your contact, you might well end (a little capriciously!) with '73 and 88' – 'Kind regards and love and kisses'! Seriously, I would contend that if everybody subscribed to the culture of friendliness and helpfulness displayed from the very first moment of contact by radio hams, mankind would be rid of all wars, and rumours of war. You see, by its very nature, ham radio recognizes no boundaries, and assiduously ignores racial and religious differences, and embodies a camaraderie almost unbelievable until you have actually experienced it at first-hand.

But to continue with those surprises ham radio can spring on you... First, I need to point out that *women* radio hams are very much in the minority (do they find all those knobs and dials off-putting?). All the more surprised was I then, when, having called 'CQ DX', the sound of a woman's voice came out of my speaker. She turned out to be in South America (I've forgotten which country). In keeping with the general informality characteristic of ham radio, we immediately exchanged first names – and *hers* was *'Dinky'!* For some reason I found the thought of someone called 'Dinky' living in South America somewhat incongruous. But there you are – that's ham radio for you!

The third surprise I recollect was sprung on me at midday, one Saturday (why do I remember it was a *Saturday?* Did the surprise I experienced etch every detail, significant or not, on my memory?) I called, 'CQ DX', and waited, and up popped a New Zealand sheep farmer: it was *midnight* for him, and he was about to go to bed! You see what I mean now about the world 'being one's oyster'.

A last thought about the appropriateness of my new

hobby, in the context of being Mary's full-time carer. It meant that, for example, having settled Mary on the bed for her afternoon rest, and confined to the house though I was, a few steps into the second bedroom, and I could soon find myself spirited away to, say, Canada, talking to a trapper in his hut on a remote mountainside in the Rockies. The circumstances being what they were, what better hobby could have 'fallen into my lap', as you might say?

Chapter 19

Back now, to 1977, the year in which Mary had her second stroke; but also the year of the OBE, and the trip to the Palace with John, Roger, and Rosemary.

Mary had spent about two months, first in hospital, and then in a Rehabilitation Unit, and had returned home to Thorn Bank towards the end of August. At the beginning of September Roger and Wendy and family paid us a flying weekend visit, arriving on the Saturday, and returning to Bath on the Sunday afternoon.

In the aftermath of their sudden departure, I had wondered when, and whether, Mary and I would, together, ever see Bath again: at the time such a journey seemed like a space odyssey merely to contemplate. Yet, as Christmas drew near, there seemed to be a desperate need for us to break out of the cribbing space of our own little house, which seemed to be shrinking by the day, and to know once more the liberating experience of new places and spaces, or the refreshment of spirit to be gained by re-visiting old and familiar ones.

I talked to our doctor about it. Was it mad, or at best plain foolhardy, even to contemplate the journey to Bath? (Roger and Wendy had already invited us for Christmas.) Greatly encouraged by the progress Mary had made, I think he was determined that we should not mark time. 'I should go,' he said, adding something to the effect that I couldn't keep Mary wrapped up in cotton wool indefinitely, that there would always be a measure of risk involved in whatever we did (or didn't) do, and what better cause to take a risk for, than Christmas with the family?

It is sad to think that nothing comes back to me of the family festivities of that Christmas: just two things - very practical matters both of them - come to mind. The one was that I failed to pack the 'Stand-Easy', a vital bit of equipment which we were hard-pressed to do without, and the other was one of those photographic snapshot-style memories - of Roger walking into our bedroom one morning near the end of our stay, and without any preamble whatever asking, 'Why don't you move to Bath?' There were some town houses about to go up, he added, virtually on their doorstep, too - so, 'What about it?'

It took hardly a second to decide, and in little over six months – on June 30th 1978 - we had taken up residence in Bath!

It is said that moving house and home is a trauma second only to bereavement. Be that as it may, I got all the reward I could possibly have hoped for just after putting Mary to bed on the very first night in our new home. Snuggling back into the pillows, she said, quietly but solemnly, 'This house shall be called "Content" '. Just one word, but it left me speechless.

One can have an oasis in time as well as space, I suppose, and like that, Mary's oasis of content lasted less than three months. As I walked into the bedroom with her breakfast one morning in September, she was in a torpid state, her features contorting rapidly in the horrifying way with which I was all too familiar, and in a matter of minutes she was deeply unconscious.

Our new doctor came, clutching her notes, and immediately ordered an ambulance. As they carried Mary out, he made an opportunity to speak to me on the side. 'You know,' he said, soberly, 'your wife is, I think, every bit as ill as she was last year.' He had no reason to think otherwise, and I had no reason to hope otherwise, as I climbed into the ambulance with Mary. The scenario was indeed a horrendously familiar one.

At the hospital she was rushed into a ward where there seemed to be an emergency team ready and waiting for her, and she soon disappeared behind the curtains, with at least half a dozen people round her bed. The place was bustling and bristling with activity, and I took heart, feeling somehow that she had fallen into good hands. I rang Roger, and a little later he joined me on the bench outside the ward. Yet another vigil had begun.

We were offered the statutory 'cup that cheers', but it would have had to have been a very special one to cheer me just then. I was discovering that my feelings and emotions, which seemed to have recovered their resilience after the move to Bath, were actually almost as played out as they had ever been. It was as though that deeper level of being which had been so totally engaged in the battle for Mary in 1977 had fallen asleep exhausted; and that, as I sat there with Roger, someone was shaking me violently, and saying, 'Come on! Wake up! It's all happening again.' Was that really what they were saying? And for a third time?

About noon, the ward sister suddenly materialized in front of us. I took a firm grip on reality then, for the first time that morning, and steeled myself for the worst. She said, 'You can come and see your wife now,' adding then, with a twinkle in her eye, 'She's a fraud, you know!' It wasn't the first time that Mary's powers of survival had educed such a response. 'She's a survivor!' I declared, as Roger and I swept through the doors into the ward, to catch our first glimpse of Mary.

Strange, isn't it? - that the drama of the hospital ward goes on behind closed curtains, and it is only when it is all over that the curtains are actually drawn apart. There was Mary, comfortably propped up against her pillows, with a smile that seemed to say, 'Fooled you again, didn't I?', all the contortion of her features gone, and looking as though nothing had happened - nothing, except, that is, that she had woken up in a hospital bed instead of her own, where she had pronounced that all was pure 'content', just weeks before.

She stayed in hospital about a fortnight, mainly so that they could thoroughly reassess her drug regimen. She was also taken to Bristol for a brain scan, which showed that the trouble had been on the left side again. However, it had had only relatively minor effects on her mobility and speech. We had been let off so lightly, compared with the previous year, but the incident had inevitably undermined the growing sense of security we had begun to feel as a result of the move to Bath.

It wasn't very long, either, before a number of practical problems began to emerge. Holloway, the road in which our new house was situated, was a steep hill, and all we could do for Mary's exercise was to walk up the road fifty yards or so (which was as much as she could manage of the hill) and back again. It got very boring for both of us, and she began to show less and less inclination to take her walk each day.

That wasn't the only thing that turned out to be different from what we had imagined. Although the house was indeed only a hundred yards ('as the crow flies') from Roger and Wendy's, there was no direct road, not even a direct footpath. It was in fact the better part of a mile to them by car!

We began to realise that, being so near and yet so far from them, we might as well be a mile or so in some other direction altogether, and living in a bungalow. The trouble was that whilst there were quite a lot of bungalows in Bath, the very *raison d'être* of most of them was vitiated by the fact that they had a flight of steps up to them, or down to them, as the case may be. Nevertheless, in a gentle sort of way, from then on, the hunt was on for a bungalow on a level site. It took us nearly four years to find one.

We moved into it in September 1982, and Mary was to spend the remaining six years of her life 'on the level' – which was so much easier for her. Well do I remember one particular morning when, up and dressed after breakfast in bed, she was taking the few steps across the hall from the bedroom to the sitting-room. 'I do like this little house,' she said. All the hassle of that second move had proved worth it...

In the meantime, life was by no means one long succession of newly unearthed snags. We began to realize the advantages of living in the West Country, within little more than an hour's journey by road (Mary's limit) from such places as the Wye Valley and Exmoor, and for six years, beginning in 1980, we ventured on annual holidays again. I say 'ventured' advisedly, because we both needed to screw up our courage to leave behind the relative security of our home base, and take up residence, with all our bits and pieces, among a group of strangers, as hotel guests. But, apart from some almost inevitable contretemps, mostly it worked out well enough. Nonetheless, in the Spring of 1980 I took myself off to the doctor's on my own behalf. I was feeling somewhat played out, and rather under the weather. The doctor was reassuring, but followed up his verdict with a suggestion which I think I must have, momentarily, totally misunderstood as raising the question of Mary's possible 'institutionalization' - the word as ugly as it is long, and the very idea of it completely unacceptable to me. Actually, what I had heard him saying was something to the effect that he could 'get Mary into a "home" ' if I wanted him to.

All I can remember then was of rounding on him fiercely, and blurting out, 'Oh, no you don't - she's mine.' Yes – those were my exact words!

A moment later, and I think we both felt more than a little embarrassed, as he went on to explain that he had meant only for a week or two, to give me a break. I calmed down after that, and said I would think about it. It was the first time anyone had suggested such a thing, and for some subtle reason or other I was suspicious of the whole idea. It seemed innocuous enough, and still does in principle, but as I came to accept the argument that I needed regular breaks from the daily round of caring, there were all sorts of ramifications and conflicts that grew out of its implementation, for both Mary and myself.

It was argued on my behalf (by others, and genuinely enough, I may say) that, apart from a few hours' break on a regular basis each week, I ought also to have an annual break of a week or two. And thus it came about that, in the autumn of 1980, a place was found for Mary in a pleasant 'Holiday Home for the Disabled', whilst I went off for three whole weeks with Clifford (my lifelong friend from schooldays) to Malawi, to stay with his doctor daughter, Vanessa. It was a wonderful time for me, but I was not at all convinced that it had been of unalloyed benefit to Mary.

The trouble was the old and agonizing one - that I just couldn't be sure that I had really 'got through to her', as I tried to set out the reasons why such arrangements were coming to be considered necessary. In consequence, they came to feel increasingly one-sided, and almost imposed on Mary. Worse still, as they became part of the fabric of our lives, and I had to agree to Mary going into hospital when there wasn't a place for her in the Holiday Home. It turned out on more than one occasion that 'hospital' meant a geriatric ward of the kind that gave me nightmares at the very thought of it. In no way was I prepared to purchase a respite from the daily round for myself, at such a price for Mary; and it was only by regularizing the arrangements with the Holiday Home (thus avoiding the need to resort to such an alternative) that I could contemplate continuing to take such breaks. Even so, as the years went by, and Mary's difficulties gradually increased, she became more reluctant to agree to going to the Home at all, even for a few hours each week, let alone for a week or two, each summer.

Once again, in retrospect, I am conscious of the terrible and seemingly inexorable erosion of the quality of life and relationships that was going on in so many little ways, under the sheer dead weight of the daily round. It was, indeed, like 'a bereavement in slow motion' as our situation had once been described to me, by a counsellor with whom I had got into conversation, over a restaurant table.

During the remaining eight years of Mary's life, there were no further major crises – that is, until the final one; and, looking back, it seems now that we had become moulded by crisis, our lives geared to it, like those who - having already been through so many battles in a war - subconsciously hold themselves permanently in readiness for the next one to break out.

That was certainly true of me, but what of Mary and her ability to live in 'The Eternal Now'? Had that changed, too? Had her capacity to live the life of 'pure spirit' been finally eroded by that daily sense of threat of still further battles to come? I just do not know, and there, it must be said, lay the sheer agony of it. What I do know is that, over those remaining months and years, her zest for many of the ordinary activities of life was certainly being eroded: like going out for a ride in the car for instance, or for that matter going anywhere at all - even for going on holiday.

Increasingly, she wanted just to sit at home and read, and to receive visitors - she never lost her zest for *that* - until what turned out to be the last few weeks of her life. Were we just holding our breath then, during those eight years, waiting for the next crisis to happen? Maybe - but when it came, it came by stealth, via the back door, so to say - neither of us for the time being recognizing it for what it was.

It was 1987, and with the benefit of yet more hindsight, it is so clear to me now that spiritually I was beginning to live from hand to mouth, and Mary's courage, though as unflinching as ever, was becoming steadily more stoical, by the day.

Neither of us had either the will or the heart to go on holiday that year, and we had the first (and the last, as it turned out) of what I called our 'nominated' holidays. The idea was to designate ten days or so as 'holiday time', when we would go out as many times as possible for rides in the car, and have as many meals out as we felt like. But, when it came to it, neither of us could muster much enthusiasm for the arrangement. It is obvious, now, that by then we were each trying to flog a very tired horse. But it wasn't obvious then.

Instead, for me, it presented itself as a sort of *accidie,* a sickness of the spirit, the onset of which filled me with little short of horror:

> Not to get used
> to the sound of your foot
> dragging reluctantly
> across the floor
> behind your walking frame,
> which you manoeuvre
> like an unwilling mule
> you are desperately trying
> to tame;

> not to get used
> to the sight of your face,
> frustration-fraught,
> as you strive for the word
> that is pikestaff-plain
> to your inner eye -
> but you struggle to say
> in vain;

not to get used
to the clutch of your hand
on my arm
as we dare the two steps
to the garden below -
a journey whose hazards
you, only, can know;

> not to get used
> to the thought
> that once was a time
> when foot followed foot
> in so graceful
> a walk,
> and word followed word
> in a torrent
> of talk,
> and arm tucked in arm
> just for love
> not support;

God!
not to get used
to these things,
I say -
not to get used to them, God,
I pray...

... words wrung out of me in July of that year.

Chapter 20

'Looking back,' said John, 'it could have been basically that Mum's immune system was failing.'

John, our older son, is an organic chemist, and in drug research. He was speaking just after Mary had died, in July 1988, and he was looking back over the first six months of that year, at a whole series of disparate and seemingly unrelated things that had been happening to her.

She had had two bouts of 'flu, one in the New Year, and another in the Spring. At the turn of the year, too, she had developed a small ulcer on her ankle, where the lower strap of her caliper had rubbed. Nothing like that had happened before, over the many years she had had to wear the caliper. The ulcer stolidly refused to heal. She had complained of a sore mouth, too. which gradually became peppered with tiny ulcers which came and went, sometimes seeming to respond to antibiotics, at other times apparently totally resistant. She had had an ingrowing toenail, too, which had repeatedly festered, which, combined with the ankle ulcer, made walking difficult and very painful at times. As if that weren't enough, she began to have 'waterworks' problems, particularly during the night. The trouble was that in the small hours this problem would often look to be mere obstinacy to me, or even mildly perverse behaviour on Mary's part, as though she no longer wanted to bother.

We had a long-established routine for night-times, which involved my being out of bed for two or three minutes only, the intention being to forestall any likelihood of an accident with the bedclothes. But one of the most painful memories I have to live with now, belongs to the small hours just two nights before Mary went into hospital for the last time. I had had to change the bed twice, and after the second time I was very tired, and very cross, and I remember saying to Mary, 'You know, you're lucky I haven't smacked you.' It was not, of course, meant seriously - simply as an indication of how I felt.

It is true that when I climbed into bed for the final time at 4 o'clock I had simmered down, and as I stroked Mary's head to get her off to sleep (whilst frantically trying to stay awake myself long enough to achieve that end!) I said to her gently, 'Let's treat the whole thing as some sort of nightmare, shall we? - and go to sleep now.' And off she went.

But when John came out with his comprehensive diagnosis, all the impatience and misguidedness of my view of events in the small hours came home to roost. Poor Mary - I had been talking about smacking her, albeit jokingly, for what had probably been a bladder infection.

It all seems so plain now as one looks back, but at the time it was by no means so. With hindsight, I might have responded so differently to those events of the first six months of 1988, and which turned out to be the last six months of Mary's life.

As it was, the bare bones of the situation were that Rosemary was scheduled to come from Germany in July, with her little boy, Kevin, to spend their summer holiday with us, and - somehow or another - I needed to get Mary into better shape, so that she could enjoy their stay; and I also needed the time to get ready for it.

Life was rapidly becoming more and more like a juggling act, but how to bring the juggling to an end tidily, with all the clubs in the air, and me desperately, still, trying to keep them there? I had arranged for the community nursing sister to call on the morning of Tuesday, June 14th, to discuss our plight. It was the day preceding all our troubles in the small hours; and such, already, was my mental state that I forgot all about the arrangement with the nurse. John our friend, and Anglican priest, was due to come on his regular monthly visit, to bring communion to us; but, less than an hour before he was due, Mary suddenly said, without warning, preamble, or explanation, that she didn't want to take communion.

It is true that, the day before, she had unaccountably turned her hairdresser away, though she, too, had come by arrangement. I could just about accept that, though even that was out of character, for Mary normally looked forward to having her hair done. But communion was a very different matter. It meant so much to us, and though I eventually succeeded in persuading her to go through with it that Tuesday, the mere fact that she had questioned it was a heartbreak; and when I answered the ring at the door as the nurse arrived to keep the appointment that I had forgotten, it was obvious to her that I had been crying.

Her first words were, 'Oh, you are in a way, aren't you? Let's come in and try and sort something out.'

And *that,* fatefully, was when it was arranged that Mary should go into hospital for a couple of weeks, to give me a chance to make ready for the holiday we were to spend with Rosemary and her little boy, and to give Mary the chance of having her festering big toe, her ulcerating ankle, and her sore mouth attended to, under one 'go', so to say.

She was to go in on the following Friday: it would be June 17th. Even then, before I could have had any inkling as to how events were to turn out, I saw it as a cruel twist of fate that it should be the very day of the 20th anniversary of Mary's first stroke. But it was to be much more than a mere coincidence of dates.

That Friday morning the ambulance arrived punctually. The lady who came once a week to do some housework was with us, a fact which led to yet another of those mental images burnt indelibly, as if by a flash bulb, onto the retina of my mind's eye. They didn't use a stretcher - the two ambulance men carried Mary out on a sort of chair. And as they turned at the foot of the front-door step, Mary looked back at our friend and waved, as cheerful as ever, and called out, 'See you again soon!' But - five weeks on - our friend was placing on Mary's grave a lovely bouquet of flowers: and on it was inscribed, simply, 'To a lovely lady.'

As the ambulance doors closed on Mary, I jumped into the car, and set off for the hospital. I wanted to be there when she arrived, so that she didn't feel alone.

The hospital was not far, little over a mile perhaps, and Mary had been there twice before, so there was just a touch of 'home from home' about it. Moreover, she wasn't acutely ill - anyway, not when she was admitted: she would be out again in a fortnight, wouldn't she? - and Rosemary and her little boy would be arriving shortly after that. We could put behind us those fraught days and nights we had just been through; there was so much to look forward to now.

There was certainly nothing depressing about the ward. On the second floor of a modern block, it was light and airy, and arranged in relatively cosy six-bed bays, all accessed from a spacious concourse, halfway along which was Sister's desk, where the nurse in charge sat, writing or consulting patients' notes; or just sitting, if ever there was the time to do simply that.

Mary was put into a corner bed alongside a large window, at the remote end of one of the bays. This was in itself both a treat and a sort of privilege, for it meant that for once she was not being treated as an emergency case, placed under the nose of the nurse in charge, where a close eye could be kept on her. It was all part of the false sense of security, though, into which we were being lulled.

Nevertheless, Mary's hospitalisation just at that time, and the manner of it, bestowed on me one great blessing which nothing could take away, not even the fatal turn that events took later on - especially, indeed, not that. Suddenly, I was free of all the hassle and worse, of caring for Mary physically, and was able, so to speak, to be truly separate from her - almost for the first time, it seemed. The all-embracing nature of the physical caring for so long had brought about a kind of fusion, and, indeed, of confusion, between our very states of being. And it was as though I had been released from the bondage that Mary's handicaps had day by day placed upon me, so that we could begin to relate to each other in a new freedom of spirit. The hours which we were to spend just being with each other, especially after that final, and eventually fatal stroke, were filled (dare I say it?) with a kind of glory not of this world. Does that seem like spiritual arrogance? I hope not, for it is a simple, objective statement of how it was for both of us - that is, until Mary began to suffer the ravages of pneumonia, and even *her* calmness of spirit was overwhelmed then.

The evening before she was admitted was spent carefully constructing a list of what needed to be attended to: the infected toe nail, the ankle ulcer, and the redesign of her caliper to avoid a recurrence; a reassessment of her walking with, perhaps, the need for some physiotherapy; and, of course, her sore mouth; a mention, too, of 'waterworks'.

It looks a formidable list now, and one which, it seems, should have given rise to more concern than it did at the time. But over-arching all at that moment was that wonderful sense of release from the grinding routine of each day, coupled with the thought that none of the items on the list compared in seriousness with a stroke in which, subconsciously, I went in daily fear on Mary's behalf.

The list went over onto the second side of an A4 sheet, and whilst they were putting Mary to bed (a little incongruously, it seemed then, as I had only recently got her up) I hove to in front of the young doctor sitting at the desk, explained who I was, and handed him the sheet. He looked a little askance at it at first, his body language definitely suggesting that I had stepped out of line. However, I went on to explain to him that I might well not be present when they came to set up Mary's notes, and that with her very real difficulties with her memory, and some difficulties with her choice of words, they might have problems. His manner changed suddenly then and, carefully tucking my list into Mary's ready-and-waiting folder, he seemed to relax, prepared to allow his features to display the look of the harassed man he really was, and happy to accept help from whatever quarter it came, as he thanked me.

That first night, I left Mary at about 8 o'clock, tucked up in bed after her supper, and settling in well, adaptive as ever to whatever was happening to her. What a blessing that always was - enabling me, as it did, to leave her without any qualms or fears on her behalf. In my 'days off' I had begun to paint, and I drove straight out into the country in search of a subject for the next day. The feeling of being released from the responsibility for Mary for the time being, and from 'the tyranny of time' in its most mundane sense, was growing by the minute, and my state of mind as I drove off must have had something in common with that of a dog who, having been straining overlong at the leash, suddenly finds itself free.

I drove to Combe Hay in the gathering dusk, and sized up a little village scene which I had had my eyes on for some time, but decided against it; on, then, to Freshford, where it took no time at all to settle for the little bridge, with the Inn beyond it. The painting would take two afternoons, and I would go whilst Mary was having her lunch, followed by her 'statutory' afternoon rest. That would be something like three hours total each day, including getting there, and getting back again.

Saturday morning dawned bright and sunny, so much like that fateful morning in 1968 when Mary's saga had begun. A post-breakfast visit found her settled in as well as I could have hoped. It was one of her more endearing traits that her abiding interest in people was such that, given a change of circumstances which produced a whole new crop of faces, she would immediately home onto a new face, hardly conscious of whether the 'place' which had produced it was a hospital or a holiday hotel. Soon she would have her favourite nurse – 'my' nurse - but it wouldn't be long before some of the nurses would have a favourite patient, too.

The ward was short-staffed, particularly, it seemed, at the weekends, when an admixture of part-time staff would appear, both nursing and ancillary, and there was certainly a dearth of doctors, except for emergencies, of course. So, even on that first Saturday morning, it became obvious that nothing very much was going to happen as far as Mary was concerned until Monday morning, when the consultant was due to make his round. Even then, of course, we weren't expecting anything dramatic, for wasn't it all something of a routine for Mary, and just a welcome break, for me?

The painting went well, and, as I had been hoping, was finished by the end of Sunday afternoon, and by the end of Monday it was mounted and framed and ready to take along the next day to show Mary, as a little diversion. Painting it had been a rare treat, free as I had been of the usual constraint on the time for which I could absent myself: I was still feeling very much like that dog at last let off its leash.

The feeling was not to last for long however, for when I arrived at the ward at 10 o'clock on the Monday morning, Mary's bed was empty, and she was nowhere to be found. An anxious enquiry on my part produced the information that she had been taken to the X-ray department. This was certainly somewhat alarming, until I found out that the X-rays were something in the nature of a routine. How long would she be? They didn't know - could be quite a long wait, they said. I asked where the X-ray department was then: at least I could sit with Mary, if she was still 'just waiting'. It was quite a walk, and when I eventually got there it was to find Mary stuck in her wheelchair on her own, in a long, bleak corridor. She was so grateful. 'I'm so glad you've come. I've had the X-rays, but I didn't know when anybody was coming to take me back again.' 'Don't worry, love - I'll sit with you till they do.' I daren't intervene and take her back myself, but the incident had stirred all my old anxieties again - of Mary caught up willy-nilly in the machinery of a large hospital - and there being little I could do to make sure things went smoothly for her.

And I didn't have to wait long for another instance to arise. It was the next day, Tuesday, when I arrived for my evening visit, complete with the Freshford painting to share with her. Supper was over, but once again there was no sign of her -

that is, not until I heard her voice (I had a job to locate where it was coming from at first) calling plaintively, 'Nurse, nurse, can you help me?' Fortunately she repeated it, and I was able to locate her then. *She was in one of the loos, wedged, as she had been years before at home, between the loo itself and the wall.* She said she had been calling for two or three minutes, but couldn't make anyone hear. These memories - of Mary a hostage to fortune in a large understaffed, overworked hospital - are such painful ones still: she seemed so forlorn, so very vulnerable.

I showed her the painting, hoping that it would bring her a breath of fresh air, there within the four walls of the hospital, and be a little reminder of the countryside we both loved so much. It is true that just for a few seconds she did respond with a touch of her old spirit. But it suddenly gave way to a bout of euphoria, and some confusion, which I did my best to 'sweep under the carpet' there and then: I just didn't want to know about it.

There was trouble right at the beginning of the next day too, concerning the ulcer on her ankle. The staff nurse in charge that morning seemed to want to put her stamp on events, and had removed the highly specialised dressing, which the community nursing sister had kept on it for months, under the instruction of our GP. I discovered it was missing when the physiotherapist was putting Mary's caliper on, prior to trying out her walking. She was a kindly, middle-aged soul, and highly experienced, who had met up with Mary several years before in one of those other hospitals. She had recognised Mary immediately, no doubt in her time having been adopted as 'my physiotherapist' - an honorary title not likely to be forgotten after being bestowed with the full weight of Mary's enthusiasm and gratitude!

I went straight off to the nurse concerned, who was sitting wearing the garment of authority as obviously as one might a new coat, all too conscious of its stiffness and straightness, and the lack of any comfortable creases in it. I gave her the benefit of the doubt. 'I'm afraid the dressing on my wife's ankle seems to have got left off this morning,' I said, ingenuously.

'It's not been forgotten,' she replied, already showing some evidence of defensive irritation at the mild criticism which she might have considered my statement implied. 'We decided to leave it off.' 'But it's had that on for months,' I remonstrated, ''and as I understand it, it is the only type of dressing likely to give the ulcer a chance to heal, even though it may take a long time.'

Her expression became more defensive, and yet at the same time more determined. Here was someone, I felt, whose actions were being dictated less by her professional judgement than by some inner need to prove something to herself. 'It's for us to make such decisions,' she said, almost curtly, and with more than a hint of impatience at the ill-informedness of the layman. *Was* I stepping out of line? - even being a little paranoic on Mary's behalf? Be that as it may: at least I was the patient's husband, and there was no-one else around who was going to fight her corner for her.

I went back and unburdened myself to our friend the physiotherapist. She said simply, but significantly, 'Leave it with me.' The next morning (Thursday) the dressing was back on again. But let it be noted that I had reason, much more serious reason, to doubt the judgement and motivation of that particular nurse later still - in the closing days of Mary's life.

Meanwhile, on the Wednesday and the Thursday afternoons, whilst Mary was resting, I did another painting, this time of the weir at Avoncliff. I planned to get it mounted and framed on the Friday morning (I knew a framer who would do it whilst I waited), ready to take to Mary on the Friday afternoon.
I remember having a lively and forward-looking conversation on the telephone with Clifford's wife Peggy that morning which did, in fact, delay my arrival at the hospital by a quarter of an hour or so. In the normal way it would have been of no consequence:

I might well have found myself waiting on a seat on the ward concourse whilst they finished getting Mary up, or whatever. And anyway, she would enjoy seeing Avoncliff again in the afternoon – or so I was thinking. Take things leisurely for once, I told myself, there really was no need for any hurry.

> But when I came,
> as usual,
> soon after breakfast-time,
> and sought you in the day-room
> in your favourite chair and place,
> I found you, yes, I *found* you,
> oh! *yes,* I found you there;
> but no-one, but *no-*one
> had found you'd lost your speech.
>
> My poor beloved
> Mary dear,
> they told me that the nurse
> who got you up that day was new -
> would not have known that you could speak
> the day before.
>
> God! the horror of it all -
> even to me you looked the same
> as I came up to your chair,
> but when I asked, 'How's you today?'
> All you could do was sit and stare,
> gesticulate at me
> and murmur,
> incoherently.
>
> To think that all in ignorance
> they'd got you up,
> and dressed you;
> put you in your wheelchair too,
> and pushed you to the sitting-room
> to wait for me to come;
> and all that time,
> my poor dear love,
> you'd failed to make them understand
> that you had had yet one more stroke.
> Oh, *God!* -
> the horror of it all.

And *that* was part of Retrospect, written a few months after Mary died.

I dashed back as fast as I dare along the concourse to the nurse in charge at the desk. 'My wife has had another stroke,' I blurted out, staying as calm as I could. 'Are you *sure?*' she asked, incredulous. But a sobering thought struck her then. She spoke slowly, working out its implications as she went along. 'It was a new nurse that got her up - she only came onto the ward this morning. It would have been the first time she had seen your wife....' She stood then, suddenly. 'I will come and see her.'

There was no doubt, of course, and Mary was taken straight back and put to bed again, whilst they called for a doctor. Once again, it seemed, just when my back was turned, Mary had suffered yet one more calamity.

Chapter 21

It was natural to be making immediate comparisons: June 17th 1968; June 21st 1977, and now June 24th 1988. To add to the dreadful familiarity of it all, even the day of the week fitted, for June 24th was also a Friday in 1977. What was it about that third week of the month of June?

But there were other more helpful and more hopeful comparisons. Mary had not lost consciousness this time: indeed, mentally, she seemed more lively than she would normally have been so early in the day, though frustration would probably be a better word than liveliness to describe her state. Even that was some comfort, compared with the awful passivity of a comatose condition. One was clutching at every straw that floated by. There seemed to be full movement in both her arms, too (there was no means of assessing her leg movement just then), and she was certainly making good use of them to express her frustration. Mary was a great communicator, and loss of the means to communicate would weigh far more heavily with her than any loss of mobility.

Altogether, the comparisons seemed reassuring, inviting the conclusion that things were not as bad as they had seemed to be, under the impact of my shock discovery. After all, she had been deeply unconscious in the autumn of 1978, and yet had survived virtually unscathed. What I was failing to do (influenced by Mary's fighting spirit, and her capacity to survive, so amply demonstrated in the past) was to make proper allowance for the fact that she was 20 years older than she had been in 1968, and 10 years older than in 1978. There was John's point too, that her immune system was probably failing. With the benefit of his hindsight, I might have realised that the real threat to her lay elsewhere, and not with the stroke at all. Tragically, such hindsight might also have benefited the nurse who was so eager to use the authority vested in her.

It was Friday afternoon, and the weekend was upon us again, when there would be the usual preponderance of part-time nursing staff on the ward, combined with a dearth of doctors. This meant that it would be Monday before any serious work on Mary's rehabilitation began - a thought born of my anxiety and impatience on her behalf, rather than any serious lacks in the actual running of the hospital, apart from those springing from a perennial shortage of funds. Fortunately - and as some sort of compensation for the enforced lull in the ward's activities over the weekend - it would be the consultant's ward round again on Monday morning, which at any rate (I thought) would get Mary off to a good start, at the beginning of the week.

The main fact to have emerged was that apart from having lost her speech again she had also lost her 'swallow reflex', which meant that she could not take solid food, and could not swallow even fluids properly. Most of my time at the hospital that weekend was spent trying to trickle water into her, whilst at home I developed a method of cooking four little plastic pots of egg custard simultaneously, in the microwave. They were perfect - and were to comprise a major item among the various foods I was to try so desperately during the next week or two to get Mary to take.

By the beginning of the week I had largely taken over her fluid intake chart. The target was a minimum of about 500ml a day, but it was rarely if ever achieved, and I was never sure how much of what I trickled into the side of her mouth as she lay back on the pillows actually went down her throat. It was *so* satisfying when I was rewarded with the sound (and sight) of a feeble gulp, which meant that the latest teaspoonful had found its way to its destination!

This task of mine was virtually a self-appointed one, the ward sister, perpetually short-staffed, only too happy to accept the help of another pair of hands more than eager to take on such a time-consuming task as trying to spoon in at least a pint of fluid a day, teaspoonful by teaspoonful.

In desperation I resorted to other methods too, including a spray intended for scent but filled with water, which I would use whenever Mary's mouth dropped open. The spray held 10ml, and I conscientiously recorded it on the fluid chart each time I needed to refill it. Mary's mouth, still ulcerated, used to get so dry and parched, and how gratefully she used to hold it open sometimes, just to receive the cooling spray! There was so little one could do for her.

So far as fluid intake was concerned, I was clearly losing the battle, and after a few days she was put on a dextrose drip.

By Tuesday, four days after the stroke, she had begun to perk up somewhat. When I arrived after breakfast she came out with the first four words she had managed to produce since having the stroke. She no sooner set eyes on me than she said simply, but very firmly, 'I wan' come home.'

It is rare indeed that something happens which makes one both very happy and very sad at the same time - it was so wonderful to hear her speak again so soon, but so terribly distressing not to be able - there and then - to fulfil her wish.

I found myself saying to her, quietly, earnestly, as though making her a solemn vow (which I was, of course), 'As soon as the nurses and the doctors have got you just well enough for you to be able to come home, and for me to be just able to look after you again - that very moment you shall come home again.'

We were looking straight at each other as I spoke, and her eyes lingered a little, looking deep into mine, as though she was drinking in what I had said, then savouring it. It was one of those occasions, all too rare for many years, when I was able to feel that I had got right through to her. She didn't attempt to say anything more just then, but I felt sure I had convinced her. In the event, however, things took a turn for the worse in a matter of days, and that little cheery call to our friend as she had left the house only ten days before, 'See you again soon,' was never to be realised. But I take comfort now from the thought that after that little exchange between us she knew that nothing would stand in the way of her return home except some quite insuperable nursing problem.

For the rest of that morning, we were both riding high in spirit - and in hope too, I believe. She loved travel and nature books, and for her birthday that year I had bought her a book called Nature of Australia, and I had taken it along, to turn the pages over with her, and to talk to her about the wonderful pictures of the flora and fauna and landscape with which it was teeming. In retrospect I see such brief times, of which there were several in the first week following the stroke, as part of the lull before the storm which was to follow; and it was to lull me into all sorts of false hopes and expectations for the course that Mary's illness would take. As we sat there, thumbing through the pages together, the atmosphere of a hospital ward fell away from us, and we could almost have been in deck chairs on a holiday beach, taking the sun.

In the middle of this John, our friend and priest, arrived, and hard on his heels came Elizabeth, another of John's parishioners, who had been visiting Mary for almost the whole time we had lived in Bath. She was a prime example of one of those who had started visiting Mary to alleviate her largely housebound state, but who had soon fallen under her spell.

Their arrival led to another of those mental images destined to stay with me for the rest of my life - of Mary holding court, so to say, with the three of us. I was so thrilled that she had formulated her first four words, and I was regaling the others with this fact, and recounting other little incidents and hopeful signs that had occurred. And, as I did so, Mary, full of spirit, teased me with jaunty 'good old me' gestures, tossing her head, and thumbing imaginary lapels. What spirit she showed that morning! *She actually sat there, four days after having a stroke, entertaining us.*

How feeble my spirit, by comparison. By lunchtime the thought of the long haul that lay ahead of us was yet again bearing down on me, grinding out of me the joy that I had experienced earlier, when I had joined Mary in her Eternal Now;

and soon I was watering the morning's brief heaven with my tears again. They were the old tired, tormented tears for the Mary of twenty years before, and for all that might have been. For one brief hour I was to 'Rage, rage, against the dying of the light,' as I scrimmaged and scrabbled for words to express the dark anger that had erupted from my depths. The words fizzled out though, the anger quickly spent, as the vision of Mary came back to me, as she had begun that morning to tackle the daunting task of working her way back to some kind of normality again for the third time in 20 years.

But when I returned mid-afternoon, it seemed that the effort she had put into it all, no doubt for the benefit of all three of us, had proved too much for her, and she had wilted like some delicate, exotic plant put out into a biting wind.

By the next morning she seemed to have largely recovered again, and during the rest of that week showed further small but significant signs of improvement. She began to say a few more words, like 'Hullo,' and 'All right' (the latter sometimes a somewhat grudging agreement to take some fluid after I had impressed on her the importance of doing so - she made it plain in her tone of voice what she thought of my persistence!).

On the Friday, a week after the stroke, I took her flowers purchased with money sent by her brother. A nurse was with us as I gave her the flowers and explained that they were from 'Eric', and she immediately turned to the nurse and said very plainly, 'My brother.' Unless one has experienced being cut off in this way from someone one is very close to, there is no knowing the music on the ear which two simple words like that can be.

In the meantime, my enthusiasm for keeping up Mary's fluid intake knew no bounds, and by the end of Saturday morning I had managed to get her to imbibe among other things a couple of fruit Yoghurts and a Gooseberry Fool (the latter one of her great favourites).

A major contingent from the family turned up at lunchtime, led by John and Roger, the idea being that subsequently we would all go out to lunch together whilst Mary was taking her rest. It became a very belated lunch, for by the time we were ready to go, Mary seemed in great distress, and it was very difficult to leave her in such a state without knowing what was troubling her. Again, there is no describing the experience unless you have been there yourself; but in the end I had to leave her, still not knowing what was wrong.

When I returned later that afternoon she seemed quite alright again, and I was greeted by a wry smile from one of the nurses. She said, simply, 'Try not to give your wife too many Gooseberry Fools in future!'

Despite my over-enthusiasm with the likes of Gooseberry Fool, Mary was reported as having had a very good sleep that night, and when I arrived after breakfast on Sunday morning she was absolutely wonderful. She was already up and dressed, and sitting in her chair, her face bright and smiling as I turned into the bay. There was, at once, a feeling of transformation in the air - as though something had 'clicked into place', so to say, and changed everything. She tried herself out on a whole lot of new words - not very successfully, but fresh effort and enthusiasm had seemingly sprung from nowhere. How well I remembered, from 20 years before,

> ... that little burble
> interspersed
> with laughter -
>
> and what matter that
> meanings sometimes went astray
> in wordless chatter? -
> what matter?
> Oh! how we met!

It is possible that the improvement she showed that Sunday morning would in any event have proved a 'flash in the pan' - that will never be known for certain - but ever since that day, and what actually transpired, I have been tempted to think tragically otherwise.

It was the weekend, remember, when the ward had largely to run itself, with little or no high-level supervision or help, except in the case of an emergency. And there was a sense in which Mary's sudden and spectacular improvement was itself a major matter, as much in need of that high-level help and advice as any life-threatening one would have been.

The nurse in charge was the one with whom I had had the brush ten days earlier, concerning the dressing on Mary's ankle ulcer. She was her usual bustling self, still needing to prove something, still anxious to exercise as conspicuously as maybe the authority vested in her. The combination of such a trait with the circumstances obtaining in the ward at a weekend was potentially a dangerous one, and even now, years on, it is difficult to persuade myself that it did not at least hasten Mary's end. There was no question of negligence, just a surfeit of enthusiasm, combined with a monolithic attitude to the wielding of vested power and authority. And, to redress the balance still further, let it be said that this particular nurse was one of the very first to come and try to comfort me, just after Mary had died.

What happened then, that fateful Sunday morning, when I had found myself so unexpectedly on Cloud Nine? The nurse's response to the sudden change in Mary was at once both dramatic and drastic. Until then, since the stroke, Mary had stayed quietly in the armchair by her bedside, out of sight and range of the hustle and bustle of the day-room, with all its comings and goings. But no sooner had the nurse recognised the improvement in Mary, than she immediately ordered her to be wheeled off to the day-room. She gave no reasons, but I would imagine that the decision was based on the belief that the response to any sign of progress should be to apply immediate pressure to produce still more. There seemed to be no recognition in such a philosophy of the need for, and the benefits of, consolidation. 'Press on regardless' was the order of the day, with no appreciation of the need to 'Hurry up slowly.'

Merely to spend the morning in the day-room, instead of by her bedside, might have made little or no difference to Mary, since I was to be with her, and we had our own ways of remaining fairly oblivious to our surroundings when we were together. It was what happened after that that did the damage, I believe. Instead of being allowed to go back to her bedside again at lunchtime, and my being able to give her some of the little pots of 'slip-down-easy' food I had been feeding her with all week, it was insisted that Mary should sit up at table with all the others, and try to eat the ordinary food as best she could - 'cruel to be kind' is as good a gloss as can be put on it. In the event, and after nearly an hour on the part of a nursing auxiliary spent trying to get even a few mouthfuls of food into Mary, they gave up, and came and asked me (I had been waiting anxiously on the concourse) if I would try.

My poor dear Mary - she was utterly exhausted by the time I was called in to help, and I could get nowhere with her, though I tried for another half hour or so. Gone was the opportunity to feed her with anything at all, for the time being - even one of those little egg custards - and I went to the nurse and said as much, at the same time reminding her that Mary rested on her bed for two hours in the afternoon, and thinking to myself as I did so of how much she would need that rest, in the circumstances. I said that I had to go and get a bite of lunch myself, and would be back again at teatime, after Mary's rest.

I was late getting back, for I had been late leaving. It didn't seem to me to matter much, as Mary would have been safely back on her bed, and resting for most of that time. Or so I thought.

But she was *not* on her bed when I got back, and not in her armchair either. I found her in a gaggle of armchairs in the day-room, engulfed in a crowd of other patients and their Sunday afternoon visitors, and in a virtually collapsed state. I asked the first nurse I could set eyes on whether Mary had had her rest. She looked blank, and said that she didn't know she had to have one. Apparently, after the exhausting time of the lunch-hour fiasco, Mary had spent the whole afternoon on her own amidst the hassle of the day-room, when she so desperately needed to be resting on her bed.

Following my intervention, they put her to bed there and then, in the early evening. But it was too late to prevent the day from becoming a catastrophe for her: simply because she had shown such singular improvement that morning, she had been kept up in the day-room for eight or nine hours, and expected to make a good showing at eating normal food again - yes, just like that. In such a situation there is no point in attempting to lay blame ('It is given only to God to know why anything happens'), and in the last analysis the nurse concerned was probably as much a victim of her own circumstances as Mary was of hers.

The remainder of that evening was spent with Mary, trying, on and off, to get a little nourishment into her, to take the place of what she had missed at lunchtime. In the end, she did have a little egg custard, and a little ice cream, and last thing I managed to get her to take a cup of Horlicks, but it amounted almost to force-feeding her. Part of the way through it, desperately anxious, I said, 'You must drink this, it is so important.' And at the very end of a disastrous day, which had begun with such promise, she gave me my reward - just four words, but spoken so plainly, 'All right, I will.'

I was not told what sort of night she had had after that terrible Sunday, but when I arrived after breakfast on the Monday morning the contrast with the previous morning was simply horrifying. Again the image of a wilting plant reared itself in my mind's eye, but added to it now was a strange air of preoccupation. As I sat chatting, trying to capture her interest, she was staring round the ward unhearing, looking first at this, then at that, sometimes as though she had never seen it before, and at other times, it seemed, with apprehension, as in a nightmare, almost as though she was expecting some strange and frightening apparition to appear. It was behaviour I had never before seen in her, and, given my long conditioning regarding even the slightest change, I was puzzled and inexplicably perturbed by it. In the middle of all this the consultant arrived, to begin his round of Mary's bay. My presence - as the only visitor so early in the day - would have made it difficult for him to ignore me anyway, but the thought of what Mary had been like only twenty-four hours before, compared with her strange mood that morning, and the bad impression it might make on him, gave me the courage I needed, and goaded me into gate-crashing his ward round.

Far from brushing me off as a mere layman (as he might well have done), he took me very seriously indeed, and for a few minutes I might well have been one of his retinue, much to the discomfiture of the nurse responsible for the misadventures of the day before. There was no way in which I could avoid all reference to what had gone on, but I kept it in as general terms as possible, describing to him in some detail how magnificent Mary had been to begin with, and how the day's regimen had exhausted her, so that, among other things, she had been unable to take as much food or fluid as she had on the Saturday. Mary had perked up with his arrival, and actually even smiled a little mischievously, like a naughty schoolgirl, as I told him of some of the difficulties I was having in getting her to take her food and drink.

He thanked me for what I was doing, asking me to 'carry on with the good work', and winding up with the practical advice that whatever I gave her should be as cold as possible, *ice*-cold if possible, since that would stimulate the swallow reflex. All this time the nurse in charge looked on, not a little out of countenance, even looking somewhat frustrated that her decisions of the day before should have been challenged, albeit obliquely, in this way.

Soon afterwards a young physiotherapist turned up, and did some passive physiotherapy with Mary, which only served to emphasise the terrible contrast with the first week, when she had walked on my arm down the concourse so that they could assess what, if anything, was wrong with her caliper. How quickly all that had changed... Then, I was worried about Mary's walking, wondering what their assessment of it would be - but that Monday morning, as the physiotherapist strove to get even the simplest of responses from Mary's legs as she lay flopped on the bed, I would have settled for that walking so thankfully - little hobble and all.

After the physiotherapist had gone Mary took a 5oz pot of yoghurt and 3fl oz of water in some sort of fashion; and later, at her lunchtime, a 6oz egg custard and a further 4oz of water. Optimistically, I duly made the corresponding entries on her fluid chart – optimistically' because I was never quite sure how much of what I managed to put into her mouth actually trickled out of the side of it again and got absorbed in her bib. These entries on the chart were beginning to relate more to my own comfort and satisfaction than to hard facts.

This particular feeding session had been difficult for both of us, and at the end of it I said to Mary that I really had to go then to get a bite of lunch for myself, or I would probably, as I put it, 'sink through the floor.' And to my amazement, she gave me a genuine little giggle, by way of a reply. She was so game.

Whenever I left her bed on the way out, I had to pass one of the men's bays, and I had become friendly with a young man whose late middle-aged father had been admitted for general nursing care following a fall downstairs. There didn't seem to have been any serious damage done, and certainly no broken bones. The two of them lived together, the father widowed, the son an only child, and he was desperately concerned about his father's fall. I remember succumbing to a tinge of envy the week before, as I compared his father's age with Mary's, and the relative lack of seriousness of his accident, compared with Mary's stroke. This had been followed by a strong sense of fellow feeling when he had told me one day that he was having great difficulty in getting his father to take his food. He was as worried about his father (unnecessarily I thought, at the time) as I was about Mary, and he began to spend as much time as I did at the hospital - which meant all his time, apart from his own meal times. The fellow feeling became mutual, and deepened steadily with the exchange of news about our respective patients whenever we met.

I asked him how his father was, and he told me that he had developed pneumonia. He was shattered, and I tried to comfort him, saying that his father was relatively young, and that modern antibiotics were so effective. He looked pathetically grateful for the encouragement, but turned away, and went back to his vigil.

Under stress, one's mind tends to be like a ship, with its watertight bulkheads that can be sealed off from one another, to prevent the contents of one compartment from invading another. Strokes were strokes, and pneumonia was pneumonia, and at least Mary hadn't got an acute illness like that to cope with - or so I thought, that Monday lunchtime.

The afternoon and evening were much the same as the morning had been. There was that same strange and peculiar air of preoccupation about Mary, as she sat in the armchair by her bed, gazing round at the other occupants of the bay, and at the comings and goings of the ward staff. It was almost as though she had just arrived and was trying to make out what it was all about: I had the greatest of difficulty in getting her to pay any attention at all to the perpetual problems of food and fluid intake.

I always went back last thing in the evening, around 9 o'clock or so, when the final round with the drugs trolley was under way, and patients were being tucked up for the night. I needed to see Mary settled, and, if possible, to know that she had gone to sleep, before I left the hospital. That was the only way I could settle down for the night myself, once I was home again.

But she was restless, and rather than continue to be a possible distraction at her bedside, I went and sat on the concourse, within sight of her, so that I would be able to tell when she had finally gone off to sleep.

It got to half past ten, and the nurses came to me then, saying that Mary would be in good hands, and virtually ordering me to go, pointing out that I myself badly needed rest. I went, reluctantly, and with heavy heart, feeling that there was something about the day, and Mary's strange behaviour, that I hadn't understood. And if I had had any thought that a night's sleep was going to change things for Mary, I was to be gravely disappointed in the morning. When I went as usual, after breakfast, it was to find her still in bed, and obviously very under the weather now. For most of the day she lay, propped up by her pillows, only occasionally opening her eyes, even whilst I was trying to feed her. A doctor who came to look at her briefly told me that the 'pots of this and that' didn't matter now, but that the fluid was important, and during the rest of that day I managed to give her around 400ml of water, and most of two ice creams. But when I left in the early evening, after ward supper-time, it was with the growing realisation that. quite apart from the aftermath of the stroke, Mary was becoming ill in quite another way.

I had my evening meal with Roger and Wendy, and shared my fears with them. I was beginning to feel that I couldn't trust my own impressions any longer, and said as much, and Roger offered to go back with me then at 9 o'clock, for my final visit of the day, so that there would be the benefit of a second opinion. It was dusk when we arrived, and the main ward lights were already dimmed for the night as we approached Mary's bay. There were no nurses about: they were down the far end of the ward, busy with the drugs trolley.

It was yet another of those moments of sheer horror, the sound and the sight of it fixed for ever on memory's ear as well as mind's eye. Long before we reached Mary's bed we could hear the awful sound of her breathing. Afterwards, we came to know that so characteristic was this sound, and so specific its diagnostic significance, that it had been given a name of its own, based on the names of two early nineteenth-century physicians, who had observed and documented it. I was blissfully ignorant of both the name and its significance, as I tore along the concourse to seek out the drugs trolley, and its attendant nurses. One of them, looking suddenly very serious, came back with me at once. She stopped for just a moment by Mary's bedside, just long enough to say, '*That* is Cheyne-Stokes breathing,' adding immediately, 'I must get a doctor.' Contrary to my ill-founded belief of only the day before - that it was something she wouldn't have to cope with - Mary, in fact, had pneumonia.

Once again a flood of questions. How long had she been like that? What difference had that made? - and how much longer might it have been before her condition was discovered, had Roger and I not gone back for my usual bedtime visit? Half an hour? An hour? What effect might that have had?

And the events of those last three days, apparently disparate and unrelated, began to fall into place, like the parts of a jigsaw that, seemingly having borne no relationship to one another, are suddenly seen to be the component parts of a coherent picture: Sunday, when the ill-considered over-enthusiasm of a nurse led to the virtual exhaustion of Mary by the end of the day; Monday, and the sudden withdrawal of Mary's energies and her relative inability to cope with her surroundings any longer; and Tuesday, the onset of rampant pneumonia.

Chapter 22

The doctor they found that Tuesday night was a young woman, who, with her limited amount of training and experience, did her best to cope with the situation in which she found herself - that of dealing with *me* at that time of night, as well as with Mary. She took me on one side, and said baldly, bleakly, 'Your wife may well not survive the night.' The conditioning of 20 years rose to the surface then, and I found myself declaring for the second time in those 20 years, and with the same vehemence that I had used in 1978, 'My wife is a *survivor*.' Together we had won so many battles, major and minor ones alike, that I had come dangerously near to believing that Mary would always survive *whatever* life flung at her.

Despite this, as Roger and I watched over her in the small hours of that night, I found myself briefly contemplating the ultimate - of Mary not surviving, and the repercussions of that for myself. There had been those who, immediately on hearing that Mary had had yet another stroke, had (in effect) whispered, 'You must let her go now, Lloyd. She has had enough, and so have you, too. It is time.'

I had understood what was being said, and that nothing but kindness had motivated it, but I well remember responding to one such firmly expressed opinion with the equally firm statement that I was 'incapable of looking in two directions at once': I knew that it would be impossible to do all that was within my power to fulfil Mary's cheery parting remark to our homely help, 'See you again soon,' and at the same time to be deliberately planning for a new life without her. Nevertheless, the seed thought had been sown, and, for a few moments, in the weariness of the night watches, it came back to haunt and (must I say it in the good, old-fashioned way?) to *tempt* me with the thought that perhaps, after all, Mary herself (if she could have done) would have said that she had had enough. For a brief instant the prospect of us both being freed from the bondage of her handicaps held me, but a mere moment later, and it came to me that, given only one wish for the rest of my life, it would most certainly have been for Mary to get well enough to be able to enjoy her life again, and for me to be able to look after her once more.

Thus did that night's vigil serve its unanticipated purpose - that of clarifying just where I stood with Mary, and that, in turn, produced a great peace of mind for me during what did, in fact, turn out to be the last days of her life.

They had given Mary an injection of antibiotic, and put her on a drip; and within a very few hours we were able to observe her breathing becoming normal again. By the morning I felt my faith in Mary completely vindicated: she had survived.

We left before breakfast, and I managed two hours sleep before going back again. Mary was up but not dressed, in her armchair by her bed, wearing that strange, 'preoccupied' look again, as though she no longer quite belonged to her surroundings. Whatever state of mind it was, it seemed to apply to everything, including my attempts to feed her, and to get fluid into her. She seemed to be barely able to attend to anything. From time to time, though, there were bouts of frustration or anger (it was impossible to tell which), fuelling the sense of a brewing crisis, despite the fact that she had survived the night. The inability to discern what was troubling her and thus to be able to do something about it was an agony in itself, which grew to fill the whole morning.

She wouldn't (or couldn't?) open her mouth even once to take any food, and there I had to leave it, at 1.30, when they came to put her back to bed. Home then, perhaps to rest a bit myself, before going back mid-afternoon.

Nothing had changed. Mary was fretful and frustrated from time to time, though she did manage to sleep a little, in between. I was desperate to get at least *some* fluid into her, and teaspoonful by teaspoonful, I did manage to spoon in about 120ml or so. But she didn't really swallow it. It just 'gurgled' down; and how much of it left again by the other side of her mouth was once more impossible to tell. The entry of 120ml on her fluid chart was so much wishing thinking, and for my own comfort, rather than as any statement of fact.

The food trolley came round, and, as other patients, able freely to express their choice in words, were being served up with hearty suppers, I fell victim for a few moments to a combination of simple envy and resentment. I imagined what it would be like to able to accept a whole plateful of food like that for Mary, and to feed it to her, as we chatted of this and that. Throughout her illnesses, I would from time to time fall for such fantasies, only to have to recognise them for what they were, and to come back with a jolt to the harsh realities again.

Recollecting, then, the advice of the consultant about ice-cold food, and with a strange feeling that the food trolley was in another world to which Mary and I no longer belonged, I went up to it and asked the nurse if I could have an ice cream for Mary, almost as though it was some special concession. But it was of no use - she took little or nothing of it, and I had to leave again, taking comfort from the fact that she had maintained a reasonably normal breathing pattern all day, but all too aware that there was something which was troubling her, the nature of which I had failed completely to understand. It was my turn, then, to be overwhelmed by frustration.

When I got back to the ward again at 9 o'clock for my bedtime visit, it was to find them changing Mary's catheter, which at some time during the day had apparently become blocked. One didn't need to look any further for the cause of poor Mary's discomfort, and one was faced with yet another example of the consequences of under-staffing. The devotion of the nurses was total, but each of them could be in one place only at a time, had only one pair of eyes with which to discern a need, and only one pair of hands with which to meet it.

Mary's breathing, though a little 'bubbly and squeaky' and a trifle fast, was still quite regular, and during that last visit for the day we had really good and meaningful eye contacts again, which were a great comfort. I felt that she had come back again, after the Monday and the Tuesday, when, it seemed, she had been so preoccupied with a far land. And that Wednesday night, July 6th, I slept in my own bed again, with a somewhat easier mind.

My recollection of the next seven days is incomplete, and blurred - probably because they were all much the same as each other, and unmarked by any further great crisis. Mary was on a continuous drip, so that for the time being there wasn't the problem of fluid intake to worry about, although, increasingly, the nursing staff were exercised to find fresh sites for the drip: Mary's arms were becoming peppered with the appearance of small bruises which each new site produced. With the aid of the little pots of egg custard and the like, I continued to try to stimulate her swallow reflex, and to get some ordinary nourishment into her, but a new and almost uncanny quality began to envelop my visits to the ward. It was to do with time, as though it was slowing up for us, as though we were in a sort of backwater all to ourselves, whilst the rest of the world went by, in its own time flux, unchanged.

One morning during that period, I arrived after breakfast to find Mary propped up in bed and looking as radiant as an angel. Indeed, I remember using some such words to describe to our GP how she was that morning. All the stress lines had left her features, and she looked suddenly thirty years younger. Strangely, it seemed, he did not appear to share my enthusiasm. I did not ask him why, and I have never followed up the matter since, but I realise now that to his medical mind it was not a good sign at all. Such is the bliss that ignorance sometimes brings.

Indeed, in the event, it all appeared to be a mere lull, which presaged yet another storm. Mary's breathing began to get very difficult again, though there was never a return to the awfulness of the Cheyne-Stokes pattern. Once again I was taken into the doctor's room, just off the ward, and told, not simply that Mary *might* not survive the night, but that she almost certainly would *not* do so.

By now my belief in her immortality was almost total, and I protested even more vehemently than I had done the week before that she was a survivor, and that I believed she would once again survive. The doctor, another woman, but a more senior one this time - probably a registrar - looked at me kindly, but her disagreement was all too evident in her features. Once again Roger and I stayed through the night with Mary, and once again she was with us still, in the morning. The doctor came early on, and with a frankly baffled look, she said, quite simply, 'You were right, weren't you?'

The pneumonia, though, was taking its toll of Mary's strength. She lay for long periods with her eyes shut, and there was little I could do except to keep vigil over her. Indeed, there seemed to be less and less that *anyone* could do for her. Not only that, but as I look back on that time I realise now, in fact, that those in overall charge of her were already beginning to shed some of the burden of treatment, on her behalf. For my part I could not (or would not?) recognise the significance of some of the things that were said and done, at the time.

For example, there was a ward round that day, and I took the opportunity to express my concern over the fact that, owing to the very great difficulty in getting her to take her pills, Mary was falling behind with her routine drugs. I was merely told that her blood pressure was not untowardly high, and so, 'not to bother her with the pills'. I distinctly remember feeling that here, at least, was something to be grateful for - that for the time being, at any rate, she was able to do without those 'sixteen-inch gun' drugs she had been taking for so long.

For some reason, too, they took her off her drip, only restoring it (I am tempted to think now) to placate the untoward anxiety which I began to show about it. Increasingly, a feeling of extemporisation crept into their treatment of Mary, decisions looking more and more *ad hoc*, and made from hour to hour, often with the appearance of quite arbitrary changes of direction. It was as though the overall battle strategy had been abandoned, and decisions relating to Mary were being taken by the field commanders.

We struggled through in this way to Saturday, July 16th, which I have good reason to remember as the day of the Hospital Fete.

I suppose that in my heart of hearts I was aware that things were getting more and more difficult, not only for Mary herself, but also for those responsible for her nursing care. Perhaps, too, the combination of these two things was making me paranoic on Mary's behalf. Subconsciously, I was refusing to recognise that the lack, now, of any hustle and bustle round her bed meant simply that there was little left that they could do for her, apart from easing any pain or discomfort; and I was beginning to transmute my fears for Mary into a silent accusation of neglect by the hospital staff - a classic psychological side-step, if ever there were one.

And what better event to feed such a misapprehension than a Hospital Fete? I had arrived back after lunch, and the fete was already in full swing on the sports field below Mary's window. The contrast between what was going on outside the window, and what was going on inside could hardly have been more dramatic, as I desperately tried, teaspoonful by teaspoonful, to get Mary to take some water. But the issue came to a head for me when I went to enter the amount on Mary's fluid chart. *It wasn't there.* I marched up to the desk and asked for it, assuming that it had been removed temporarily for entering the record in Mary's notes. The nurse didn't have it, either. 'Where is it?', I asked. She didn't know, and whoever was responsible for making out another one was at the fete, she said, but she promised me she would get a new one made out, directly they were back.

I am sure now, as I think about it again, that deep down I knew that we were playing a game of charades - going through the motions on my behalf - and that really the time and the need for records of fluid intake were past, for Mary. But I was still not able to admit any such thing to myself, and grudgingly, and somewhat ungraciously, I accepted the promise of a new chart, as soon as it could be arranged. It duly appeared at the foot of Mary's bed in the early evening, and I, in my turn, went through the motions of entering the minuscule amount of water which I had been able to get Mary to take.

This episode of the fluid intake chart may well have set the stage for the next, and fateful day of Sunday, July 17th, exactly one month after Mary had been admitted to hospital. I had not long arrived in the ward when I was approached by a staff nurse whom I had never seen before. I was immediately struck by the fact that her appearance was belied by her manner. She was probably in her middle 30s, a tall and very well-built woman - taller than me, and certainly weighing a lot more - but she could not have been kinder or gentler in her manner, and in the way she treated me. I sensed at once that the kindness and the gentleness were born of a very great deal of understanding of where I stood that morning, and of the dire need to fulfil the task which she had been given - nothing less than convincing me that Mary *was* dying - indeed, that she would die within the next 24 hours.

Why was it that I didn't protest any longer that Mary was a survivor? It wasn't, I am quite sure, simply because the nurse had convinced me, out of her expertise, and with the weight of her experience, that there was no gainsaying the matter this time. No, it wasn't that. In fact, and on the contrary, it was because she hadn't spoken merely in such terms that I was able to take from her what I had been unable to accept from the two doctors, earlier. Although she spoke unemotionally, and, to a casual observer, even in a very matter-of-fact manner, I knew that in her spirit she had joined me, where I was, so that, in that moment of truth, I was not alone. It was *not* 'all in a day's work for her': for those few minutes she was with me, in my agony.

Thus it was that she enabled me to find myself again, and to be able to pick up the phone, and in a straightforward way to appraise the family and a handful of close friends of the situation. All the family in turn came to the hospital that afternoon, and Roger brought a little bag of sandwiches for us both, so that we could stay on through the night, without having to seek out food for ourselves. I had been told by the nurse that Mary would not be in any appreciable pain, and though I asked no questions, I presumed that that meant a modicum of morphine, to ease her into that milieu of the spirit where she would be free at last of the burden of all her disabilities.

The nurses had settled Mary down for the night, and she was as comfortable as they could make her. Her breathing was regular, but rapid, and terribly fluid-laden. They would come during the night to relieve that, with a sort of aspirator, wheeled in on a trolley - a contraption which in that situation seemed little short of fiendish. It seemed so clumsy and so crude, to have Mary subjected to its attentions in the small hours when, in another order, and some other dispensation, she could have been tucked up in her own bed at home, and sound asleep: it all began to take on the quality of nightmare.

Her eyes were closed, either in light and restless sleep, or a semiconscious condition - it was impossible to tell which - and we settled down as best we could for the long vigil. There was no question of sleep, except for the brief snatches to which tired bodies and exhausted spirits would inevitably succumb for a few minutes, from time to time. I woke from these with a sense of disbelief; and as the reality was borne in upon me anew, it brought with it the awful feeling of having abandoned Mary to her fate, albeit briefly, and out of sheer frailty of mind and body ('Could you not watch with me one hour?').

There was no way of knowing whether Mary was conscious of our physical presence in those last hours of her life, but I believe that even if she was unaware of us through the agency of her five senses, her spirit, for sure, would have known that there were those who loved and cherished her, keeping vigil with her.

She managed to soldier on into the dawn, and there followed an uneasy and restless two or three hours before the nurses suddenly appeared in strength round her bed, soon after 7am, ostensibly to wash and tidy her, and sort out the bed. They shooed us off to the day room whilst this was going on, but we had been there hardly a moment when one of them came hurrying back to us. 'Come quickly!' was all she said.

We ran along the concourse to the bay where Mary was, and arrived at her bedside just seconds before she died. I cannot use the usual cliché, and say that she died totally peacefully, for it seemed to me that right until her very last breath she struggled for the oxygen which would keep her alive, and amongst us still.

As that last breath left her body, I gently closed her unseeing eyes, and left her bedside immediately.

I had seen many dead people in my life, but it so happened that never before had I seen anyone actually die. Why did Mary have to be the very first? - Mary, the one with whom my whole life had become so inextricably involved, through daily and hourly attendance upon her.

That last gasp of hers might very well have been seen to be my last gasp, too. For I also had to die - to the old life, the life in which my very identity had become so largely vested in Mary's care. But it was to prove to be a lingering, and very painful death for me.

Chapter 23

Let me
At least take comfort
From my tears -
That there is yet left
Enough of me
To weep,
Not simply, now, for loss of you,
But for a myriad
Unacknowledged pains
Deep buried
By the wayside
Of our journey
Through those stricken years.

"I weep,
Therefore I am" -
Thus do I reassure myself
That desolation
Has not laid final waste
To me;
Yet
On that journey
Did I, like those
In heat of battle smitten,
Soldier on,
Unaware
That when the strife was at an end
I would have mortal wounds
To tend.

Weep then,
Scan the bleak landscape
Of those fraught, fateful years,
Drink the full draught
Of unassimilated pain,
And the then unshed,
Countless tears.

I had joined the U3A, 'The University of the Third Age' – at the time a relatively new, but rapidly growing organization for the over-50s, providing study and discussion groups on all sorts of topics. I had already joined a painting class run by a private art school, but my family thought that a few more regular activities would provide a framework round which I could begin to build a new life. I joined the Current Affairs Group, which met in a room let to the organization by a small convent. It was my very first attendance.

The room was modern, and pleasantly furnished, and all round it were large wall posters, excellent reproductions of colour photographs, each of which had been given an appropriate caption. One was particularly memorable. It was a photograph taken from inside a sea cavern, looking out onto a sunlit sea. The blackness of the cave walls served only to enhance the brilliance of the sunlit scene beyond. The caption ran something like, 'Beyond the darkness, the light'. At that moment the words struck a deep chord in me, as did the words on a number of the other posters. I heard little of the discussion that morning, far too busy with my own current affairs.

We had been served with coffee and biscuits by the nuns, and on the way out at the end of the morning, I passed the open kitchen door, and could see one of them washing up. She happened to look my way, and acting on an impulse I began, falteringly, and near to tears, to say how much I had appreciated the posters. She asked me then, quite simply, what was troubling me, and I told her in a sentence or two of Mary, and of her death eight months before, and found myself saying what I had said so many times during that period - that I had not only lost Mary, but my whole role in life as well.

And what did she do? Quote me a text? (you know – 'Blessed are they that mourn', and so on). Preach me a neat little homily? No - not at all! She flung her arms round me, gave me a whacking kiss on the lips, and said, simply, 'What you need is a little love.' The dammed-up tears, which 'a little love' had released in me, began to flow then, and as I tried to hide them she said, 'Jesus wept - please don't be afraid or ashamed to cry.'

She had just time to ask me my name, and to say that she would remember me in her prayers, before I turned and fled. I needed to get home as quickly as possible, and drove straight off, despite the tears: they lasted the two or three-mile journey home, and for a full hour after it. It was during that time that 'Let me at least take comfort...' was written. Thus began a process of catharsis - of drinking 'the full draught of unassimilated pain'.

Looking back, I realise now that the process had actually begun within days of Mary's dying, though many things, sad to say, had conspired to interrupt it. And it had started at the right place: at the very beginning of it all, with the pain - unassimilated even after twenty years - of finding Mary unconscious, in a coma, as I clutched the poem which was my vision of the future for us, but which was never to be realised - not in this life, anyway.

It had been a devastating moment, spiritually as well as emotionally, but as I read the poem over and over again in the immediate aftermath of Mary's death, it took on an entirely new aspect and relevance.

> What does it matter,
> but that I meet thee now?
> my being with yours,
> finding each other,
> and ourselves, anew,
> after the long day's heat...

It came to me then that 'the long day's heat' was no longer the arduous time we had spent bringing up a family, but could be seen as those twenty years in which both of us had been in bondage to Mary's disabilities. Released from that bondage at the moment of Mary's death, our two spirits were no longer inextricably tied in with each other: could not we, once more two separate beings, find each other, and ourselves, anew, in that separateness? There was fresh hope in such a thought, and it quickly became linked with another insight I had had.

I had begun to suffer feelings of panic from time to time - feelings that the situation was intolerable, that Mary's death was something beyond my capacity to bear. These feelings mostly arose soon after putting out the light at night, or in the small hours, if I should happen to wake then. Well do I remember the prayer that I would use at such a moment, so near was it to the limit of what I could stand: 'Lord help me to be just able to bear this unbearable thing, just able to tolerate this intolerable thing, just able to suffer this insufferable thing.' To be just able to cope would be miracle enough, and there seemed to be no point in praying for more.

There came a night when, only half awake, I thought for a moment that I could hear Mary breathing beside me. I woke myself right up then, and in the panic that followed I started in my usual fashion to walk through the house from room to room, in a state of utmost agitation. Suddenly, I remembered that on that particular night, before going to bed, I had loaded both the washing machine and the dishwasher, and set them going; and by the time the panic was upon me, they had completed their allotted tasks. I made a profound discovery then - which was simply to divert the otherwise uncontrollable, volcanic energy of the panic into a practical activity. I emptied the dishwasher, deliberately taking my time over it, putting each separate piece away carefully, where it belonged. Then I did the same thing with the washing, slowly folding each article, making neat piles, and finally stowing them away in the airing cupboard. All this took quite a long time, and I followed it by stripping my bed completely, and remaking it with fresh linen: I was no stranger to changing beds in the small hours. And at the end of all this the panic had subsided. Moreover, as a bonus, I had a nice, freshly-made bed to climb back into, when I was ready. Another thing I was prompted to do that night before finally going back to bed, was to put the front door lock on the latch, so that anyone could let themselves in, if need be. This also had a calming effect: I felt less trapped inside the house, and thus less trapped inside myself.

Somehow or other, I had used the panic to do a number of jobs which, if I had got up at the normal time and tackled them, would have seemed really irksome. Their very irksomeness had completely vanished in the process - surely, good out of evil! As, at long last, I made to get into bed again, I became aware of how thin the ice had been that I had skated on that night.

In fact, in quieter moments, I began to see the panics as nothing less than a crisis of identity. For years my life had been totally taken up with caring for Mary in her state of disability and all that stemmed from it. But now the disabilities were no more, and the life which I had built round them had no reality any longer, either. Yet, I was still trying to hang on to that identity, mourning its very loss, even though Mary's life, and her disabilities with it, was at an end. As this new-found separateness was borne in upon me, I began to realise that Mary certainly would not want me to go on clinging to that old life, or the identity associated with it; less still, to mourn its passing.

I had to start again, becoming nothing less than an entirely 'new man'; but I was to discover that that was to be no easy matter. Many times there were tantalising glimpses of what it would be like when I had achieved it, but, in Harry Williams' words, 'It hurts when the manacles which chain us to the past are broken' - I had to suffer the pain of dying to the old self, and then of being born into that new self which was waiting. I believe now that those tantalising glimpses were gifts of grace, when, for a moment, a minute or two, an hour, or blessedly but rarely for most of a day it was given to me to know, with Julian of Norwich, that 'all shall be well, and all manner of thing shall be well'.

In the meantime, wandering as in a maze through the wilderness of emotions that we call bereavement, I knew for the first time the depths plumbed by the experience. Never again could I watch, detached, as a reporter on the 'box', concerned only with the newsworthiness of a story, blandly asked someone what it felt like to have just lost a spouse or a child in some tragedy or other, from a rare illness to a motorway pile-up, or a terrorist bomb. News-gathering can be almost as cruel as the event itself, and perhaps at its most cruel in its dealings with those just bereaved. I used to wonder what I would have answered when asked 'what it felt like'. Whatever the answer might have been, I hope that it would have created an awareness that reporters push their way into places where angels fear to tread, and would have made the offender think more than twice in the future, before posing such a question again.

As I look back, with insight born of hindsight, I realise that there were key moments, and key experiences, some of them very early on indeed, which, if I had properly grasped their significance at the time, would have greatly shortened the length of my journey through that wilderness. There was, for instance, Neville Ward's book, 'Friday Afternoon', to which I was introduced within weeks of Mary's death. The book is a series of meditations on Christ's last words from the cross, and deals in great depth and with great sensitivity with the experience of failure, and loss, and in particular, bereavement. I remember reading the book almost at a single sitting; and well on into it, inconspicuously embedded in the text, there was a brief paragraph which spoke volumes to me, rang all sorts of bells with me, but I seemed powerless at the time to take it to heart, and to act upon it.

Says Neville Ward: 'There is a better and a worse way of dealing with disaster. However great the good of which life has been robbed by it, *new good* [italics mine] begins to be made if we choose the better way. There are powerful kinds of good that can come into life only where something has gone terribly wrong; it just happens to be one aspect of the composition of things. There is an Old Testament prayer, "It is good for me that I have been in trouble, that I may learn thy law". Neville Ward comments: 'There is more in it than that our mistakes at any rate serve to point out and underline the right way. The writer may have discovered this subtle principle of man's emotional and spiritual life: *the curious bonus attached to good that is erected forthwith on the actual site of failure and loss.* [Italics again mine.] There were many other passages which, again with George Fox, I knew 'spoke to my condition', but which I seemed powerless to act upon at the time. There was a quotation from Gabriel Marcel's book, Homo Victor: 'It is never a simple return to the *status quo*, a simple return to our being, it is that and much more, and even the contrary of that: an undreamed-of promotion, a transfiguration.' Neville Ward goes on later to comment, 'Another word for this is "resurrection".'

And there was that other wonderful thought in 'Friday Afternoon', 'Bereavement is loving in a new key.' But I just couldn't cope with that either, at the time.

However, there was one experience – and a crucial one - to which I *was* able to respond, and which has meant so much to me, ever since. Five days after Mary's funeral, and with a thunderstorm threatening, I threw my painting gear into the car, and drove off into the country. I had no idea whether I would ever want to paint again, or, for that matter, whether I would still be able to; and, symbolically enough, I had no idea where I was going, either.

The pending storm added its own urgency - I had to be quick in my choice of subject, if there was to be any time at all to paint it. My unconscious was in the driving seat, and I had gone barely two miles along a favourite lane, hardly wide enough for two cars, before coming to a natural lay-by at the entrance to a field. I pulled in, and switched the engine off, and sat, immobile and irresolute, watching the gathering storm, and beginning to feel that it had been a mistake even to contemplate putting the matter to the test so soon after Mary's death. It was precisely the kind of mood in which, in the tradition of the countryside, one felt the simple need of a five-barred gate to lean upon - and I climbed out of the car and did just that.

From the distance came the lazy sound of a tractor trundling back and forth, lulling me into the state where the mind is quiescent, and the imagination alert. In the foreground was the golden ochre of a newly-harvested field, and beyond that the burnt sienna of an older stubble, beyond that again field upon field vanishing into the storm haze. And over it all was the lowering sky, its threat alleviated by a single strip of blue.

It was that patch of blue, more than anything else, which sent me scurrying back to the car for my painting gear. For me, at that moment, it spelled the promise at the heart of the storm, and before the storm burst, soft pastels had captured the sky and the patch of blue, together with the far and middle distance. There, was a line of dense woodland, 'blacker than black' and impenetrable, under its branches; but immediately below it, almost lurid in the storm light, was the brilliant yellow ochre of a ripe, and as yet un-mowed cornfield.

The storm was short and sharp, and the painting was finished within a couple of hours of first taking up position at the five-barred gate. Despite the initial and grave misgivings, it had - in the event - been painted with a sense of utmost urgency; and though there would be no Mary to whom I could take it home, as I had done all my other paintings, there was a distinct feeling that, unlike all the others, she had actually *participated* in this one. The painting spoke of the richness of the harvest, even under the threat of storm, and as it came into being on the painting board, it spoke to me, above all else, of the richness of the harvest of love in Mary's life, the sense of which no storm could take away, even the storm of bereavement. And all who have shared this painting with me, including many who knew nothing of the circumstances in which it was painted, have said that there was something special about it, and I do believe that *that* was because Mary had actually taken part in it. (It is reproduced in Plate 18.)

There were other occasions when Mary seems to have intervened. I am not speaking of visions or voices outside myself, but of mental images, and thoughts, and dreams, which were of such a nature, and so utterly contrary to my prevailing mood, that it would be quite impossible to attribute them to any kind of wish fulfilment.

There was the time when I had been visiting Mary's grave, some months after she had died. It was a dreadful, late-November afternoon, murky and misty, as I turned away from the grave, distraught, the tears welling. Suddenly, into my mind, and without anything at all to prompt them, came the words, 'Come Lloyd! Come Lloyd! It's not *that* bad.' Had there been anyone with me, who knew me at all well, they would have deemed my mood just then as inconsolable, and would certainly not have had the courage to address me in any such words. In any case, once the words had injected themselves into my mind I associated them immediately, and unquestioningly, with Mary.

Again, one day, as I was arranging flowers on Mary's grave, I was thinking of how, through the agency of my own two hands, two feet, eyes, ears and speech, Mary could continue to 'dwell among us', and in the very thought itself I felt I was making the offering of my being, to Mary. Suddenly then, in my inner eye, I saw her so plainly - no vision this, just a mental image. She was sitting so quietly, deeply pensive, as though pondering the significance of the thought I had shared with her. She seemed to me like a child wanting to say 'Thank you' for a special present, but unable to find the words. Then, immediately afterwards, I was possessed of another thought - that in that moment there had been a giving of each to the other, anew. But even this second thought was overtaken, as, in my mind's eye, I contemplated the serenity of Mary's appearance, and it came to me that, above all else, it had been a vision of the life of pure spirit.

There was a similar moment, too, and yet again it followed a visit to Mary's grave. Indeed, my mind was still juggling with a welter of words as I went and sat on a seat that looked past the grave to the beautiful countryside beyond; and suddenly, into the middle of my words came others, from elsewhere. Once more, they were foreign to my flow of thought, and once more, without question, I associated them with Mary: 'There is so much to look forward to, Lloyd, and not back.'

The words were so totally out of kilter with my own, and they certainly unsettled me. I had to go on with my life - I knew that; but I would take serious account thenceforward of what Mary had seemed to be saying to me - that it must be not merely to indulge myself, living in the past: it must be a clearing of the decks, in fact, for all that Mary was wanting me to look forward to. And I felt, in that moment, that she *didn't* mean just in this life.

The dangers involved in our dealings with the past are well described by Christopher Bryant in 'Jung and the Christian Way', when he speaks of 'a tendency to get bogged down in the past, to cling to it for dear life and so to be hindered from grasping the opportunities and embracing the tasks of the present. We cling in nostalgia to a happiness or security we once enjoyed and long for its recovery.'

This, one might feel, is obvious enough, but with subtle insight he adds, 'Paradoxically we cling not only to good and happy memories but to painful and humiliating ones.... experiences of inconsolable grief, experiences so full of anguish as to be repressed and forgotten, but live on in the unconscious like an invisible cancer, consuming psychic energy and sapping the individual's ability to face the present and the future.' Bryant goes on, 'The practice of thanksgiving' [which he has defined earlier as 'the acknowledgement of God in the awareness of the good and helpful factors in life'] 'fosters a healthy detachment from, a letting go of the past, for it links past experience with the present reality of God.... Where the individual has a genuine faith the painful wounds of the past *once recalled* [italics mine] can be healed by thanksgiving. For thanksgiving declares the believer's faith in God's infinitely resourceful presence even within evil, mitigating its effects and bringing good out of it.' 'Indeed,' adds Bryant, 'the power of God to redeem and draw good out of evil is at the very heart of the gospel.... Thanksgiving cannot of course change the past, but it can change its effects, *through faith and memory working together,* [italics again mine] by bringing the past into the present of God, where old wounds can be healed and life imprisoned can be set free.'

That, indeed, was the task to which I had set my hand.

Chapter 24

There was much inevitable self-scrutiny: 'ifs' and 'buts', and 'whys' and 'wherefores' and 'if onlys' - the 'Ash Wednesday' experience, as I came to think of it. I had joined the sad company of those who, in T S Eliot's words, 'walk in noise ... torn on the horn between season and season, time on time, between hour and hour, word and word, power and power', 'wavering between the profit and the loss', until I longed to 'forget these matters that with myself I too much discuss, too much explain'; longed to 'care and not to care ... to sit still'.

Of course there was some distortion of perspective in all this, and there were those among family and friends who did their best to straighten me out. Our vicar, commenting on what he called 'the remarkable fact of Mary's survival for 20 years', said that it probably had more than a little to do with the fact that she did feel so secure - and that was some comfort to me. And he added that perhaps she did have some inkling, too, of what her loss would mean to me, and that she soldiered on for as long as she did, in part, at any rate, to spare me that pain for as long as she could, even though it prolonged her suffering in this life. Certainly, that would have been very much in keeping with Mary's character to do just that.

And John, our older son, in the immediate agonising aftermath of Mary's death, pointed out that in whatever ways my care of her might have fallen short, what I had done for her had fulfilled my main hope - that she should never have to be institutionalised. I do realise that if that had had to be, it would have broken Mary's heart; and it would certainly have broken mine. As John pointed out so wisely, that implied my surviving her, which in turn meant that I had to accept the pain of bereavement as the price I would eventually have to pay, if Mary was to be spared that fate. But there was another, and more mystical comfort which I was given at that time.

Immediately after Mary's death, I turned to Harry Williams yet again, and re-read his little book, 'Becoming What I Am'. It was at the time of the worst of my self-doubting, and just when I was becoming aware that I had an identity crisis on my hands. Harry Williams is so good at advocating approaching God just as you are - no airs or graces, no pretensions: '...wherever I am, or whatever I am doing, whether I feel tired or excited, angry or amused, a bundle of nerves or calm and quiet, miserable or happy, optimistic or in despair, let me see that all I have to do is to turn simply to God and say "Hello, it is me." And when you can do simply that,' Harry Williams says so characteristically, 'there is joy in heaven among the angels of God.' But, cryptically, he goes on then to say that 'Hello, it is me' is an *answer* to a prayer as well as a prayer itself.

Now, there came one of those distraught days, when whatever I had done for Mary seemed to me to have dwindled into near-insignificance compared with what I might have done; and, close to despair, I took Harry Williams' advice, and simply said, 'Hello God, this is me.'

What happened then was a startling confirmation that 'Hello, it is me' was a prayer capable of providing its own answer. For I had hardly uttered it when I had a seemingly totally incongruous mental vision of Mary and me on holiday in Abergavenny in 1986, the last proper holiday we had.

For the first time, in a flash of awareness, I saw the risks involved (albeit unwittingly) in such an undertaking. Quite apart from managing the ordinary mechanics of living in a hotel, and away from our home base, I saw plainly the hazards of being out and about with someone as vulnerable as Mary was, perhaps on a lonely mountain road, or a mile or two into a dense woodland, on a single-width cart track miles from anywhere. Foolhardy? I don't know. What I do know, is that in that moment after I had said, in all the wretchedness of self-doubt, 'Hello, God, it is me,' it was as though, in that vision, God was saying to me, 'Yes, this is you, but that was you, too.'

Come to think of it, the test of any deep relationship should be the possibility of being able to say 'Hello, this is me,' -

warts and all, no pretences, airs or graces - and to find oneself accepted.

But I was not without my 'comforters' of the ilk that offered Job the benefit of their wisdom - all good-intentioned, but unaware of the wilderness I still inhabited, most of the time.

>   Many said
>   that it was over now;
>   others, that I should be thankful;
>   a few (I fear) that it was simply
>   God's will that had been done,
>   (like the hurricane last year) -
>   but I knew the life, which,
>   in my heart,
>   had only just begun.
>
>   Death gave it birth: your passing
>   the cosmic contraction
>   that propelled me
>   on the headfirst journey
>   through the tortuous tunnel
>   of fierce disbelief -
>   with pain beyond thought -
>   into a future
>   furnished
>   with an empty chair,
>   an empty bed,
>   coats still hanging, ready,
>   in the hall;
>   your place at table vacant,
>   meal on meal.
>
>   Morning upon morning,
>   listless and leaden-eyed
>   I lie,
>   hoping the world
>   will pass me by.
>   What chance of that? -
>   the harsh light of reality
>   leaves no margin
>   for the re-interpretation
>   of events;
>   better to weep my way
>   into the crevices
>   of yet another day -
>   others, perchance,

will neither find
nor even seek me
there.

But -
tread warily;
at every twist and turn
of time
and space
there lies in wait
some devastating evocation
of the past -
beyond tears:
from a paper scrap,
with frail words
traced by that rebellious hand,
to your first caliper, lurking
in the dark depths
of a wardrobe
where I sought
a long-lost pair of shoes.
(God alone knows
the volumes spoken
by a piece of bent iron
fitted with a leather thong
and metal peg -
the vision, too,
that it evoked
of you,
with your strapped leg
and walking frame,
and those first faltering steps -
God, indeed, alone knows
the courage that you showed.)

Beyond tears, yes,
but weep noneless;
grief's work
will not be done,
till grief itself
dies, in childbirth -
its progeny hope's glimmer,
lighting the darkest recess
of a breaking heart.

About the time that was written (some four months after Mary had died) I had 'flu very badly, and woke in the small hours to find that I had lost my voice completely. It was the second time this had happened - the first time had been within a few days of Mary's dying, the result of talking for hours on end. It was almost to be expected in those circumstances, and in a different state of mind I might even have welcomed it as a valid excuse for stopping talking altogether, for the time being.

If that thought had been in even the back of my mind, it was to vanish altogether with the disappearance of my voice. As I tried to speak, another altogether different kind of panic struck me - totally unanticipated - as it came to me that this was what Mary had had to put up with during the last few weeks of her life, yet with no sign even of frustration, let alone panic. It was another of those 'unacknowledged pains', buried deep in my unconscious, and erupting with terrifying power as I identified with Mary's plight, in the midst of my own.

That second time it happened, when I had 'flu, was to be the most horrendous manifestation I had ever experienced of the eruptive power which lies beneath a repressed and undigested emotion. Everything conspired to make it so: the fact that it happened in the small hours, when life is at its lowest ebb; that I was on my own, and, worse still, at such an hour unable to call on any human help, even on the telephone. Who could I ring up at three in the morning, to whisper as best I could that 'flu had caused me to lose my voice, and that I was in a state of panic in consequence? How to explain, at such a time of night, the dread which had been rammed into the deep hold of my mind, and had had the hatches battened down on it? For four hours, and keeping as still as possible, I repeated over and over and over in my mind, hundreds of times, a little 'dart prayer' of Julian of Norwich, which, fortunately for me, I had come across a few months earlier, and had made use of, in calmer waters: 'God, of your goodness, give me yourself, for you are enough for me'.

It was the thought, 'you are enough', that I clung to so desperately until the dawn, when I felt that I could in good conscience ring someone up.

After the 'flu, insomnia (which had never been far away) and a month on Temazipam, prescribed by a junior doctor. The drug was little or no use against the effects of the depression left behind by the 'flu, which was almost the last straw, but it yielded an hour or two of fitful unconsciousness each night, desperately needed, if only to separate one day from the next. Without that brief respite from my own company, I felt I would go under altogether.

At the end of the month I went to renew the prescription, and happened upon the senior partner this time. 'You know, Temazipam is seriously addictive,' he said, adding immediately, 'You ought to get off it as quickly as possible.' 'Then don't give me the prescription. I'll have to find a way of doing without it, won't I?' I responded, but with no hope or expectation of how I could do so.

We were almost into December, and the run-up to Christmas. Everything was beginning to pile up now, with a speed and an inexorability which was terrifying. Awful though it is to admit, even without the complications of the 'flu and insomnia I had been dreading this first Christmas without Mary. I believe that is a common experience in bereavement. There were the cards, too - we usually sent over a hundred of them - and the presents for family and friends. Such was the malaise of spirit that had come upon me so catastrophically, that if I could have done, I would have wiped Christmas off the calendar for 1988.

The most immediate problem was making up my mind about going to Rosemary in Germany for the holiday period. In the state I was in, boarding a plane for Stuttgart felt as remotely possible as boarding a spacecraft for the Moon. Moreover, doing without Temazipam was proving to be easier said than done, and virtually sleepless nights were not making decision-taking any easier. Nevertheless, the decision was made to go to Germany, aided and abetted by the community nursing sister who had been involved in getting Mary into hospital in June, and who had held a watching brief over me ever since

Mary had died. How important that lifeline was at that time, and how much more important than whole bottlesful of Temazipam!

Christmas in Germany brought its own problems, some of them anticipated, some not. There was the matter of sheer distance, the sense of having travelled too far from Mary (though quite irrational, for she was dead now). The sense of isolation was increased in a quite unanticipated manner, too. In one's own country one is linked with everyone around one by the mother tongue, enjoying the subconscious experience of belonging, of being the visible part of one's roots buried below, in one's native 'soil'. But, in Rosemary's home, as the usual stream of Christmas visitors came and went, my tourist German, sufficient to get me by in a shop, or when asking for the bus station, was quite inadequate to enable me to feel part of the festivities, as the conversation ebbed and flowed over the crumbling Stolle; and I began to feel more and more excluded from the company. Rosemary, of course, could do nothing about it: the visitors were more often than not casual ones, who spoke little or no English.

It got worse, with the sense that the family was scattered, with some of it on holiday in France, some of it in Germany, and some of it still back in England. I began to feel that I would have to cut my visit short, and go home: if one was to be lonely, better to be lonely among familiar things. Instead, however, I went out and used what little German I had to buy a short-wave radio, so that I could at least listen to the BBC World Service.

It was January 2nd, and by sheer serendipity (*was* it merely that?) almost the first thing I heard was a five-minute New Year's religious broadcast. The speaker read a passage from 'The Shaking of the Foundations', by the American existential theologian, Paul Tillich. I had the book at home, but hadn't looked at it in years.

'Nothing is more surprising than the rise of the new within ourselves. We do not foresee or observe its growth. We do not try to produce it by the strength of our will, by the power of our emotion, or by the clarity of our intellect. On the contrary, we feel that by trying to produce it we prevent its coming. By trying, we would produce the old in the power of the old, but not the new in the power of the new. The new being is born in us, just when we least believe in it. It appears in remote corners of our souls which we have neglected for a long time. It opens up deep levels of our personality which had been shut out by old decisions and old exclusions. It shows a way where there was no way before. It liberates us from the tragedy of having to decide and having to exclude, because it is given before any decision. Suddenly we notice it within us! The new which we sought and longed for comes to us in the moment in which we lose hope of ever finding it... It appears when and where it chooses. We cannot force it, and we cannot calculate it. Readiness is the only condition for it; and readiness means that the former things have become old and that they are driving us into the destruction of our souls just when we are trying most to save what we think can be saved of the old.'

I lapped up the words like a thirsty dog who had come upon a stream of clear mountain water after a long and arid trek: 'The new being is born in us, just when we least believe in it ... It shows a way where there was no way before ... The new which we sought and longed for comes to us in the moment in which we lose hope of ever finding it ... Readiness is the only condition ... and readiness means that the former things have become old and that they are driving us to the destruction of our souls just when we are trying most to save what we think can be saved from the old....'

That, surely, meant that I was as ready as anyone could be. As I thought about it a deep and wide and blessed assurance flooded over me - that this was the truth for me, the truth which, so to say, had been knocking at the door for so long, trying to get me simply to open to it. There was no clutching at straws, nothing of that feeling that here at least and at last was a faint flicker of truth which, if I clung to it hard enough, might stay, and perhaps even grow. No, there was nothing partial about the truth the words were presenting, it was truth seen in the broad, as though scales had suddenly fallen from my eyes. It was like someone switching on the light in a darkened room - there it all was, an orderly place, instead of the chaos that I had been imagining. It was truth 'in the round' - truth that was, as I like to put it, completely 'self-validating'.

I could hardly wait to get home. I knew that apart from 'The Shaking of the Foundations' there were at least three other books by Paul Tillich, waiting on my shelves. Even their titles spoke volumes! – 'The New Being', 'The Eternal Now', and 'The Courage To Be'. I began to devour them, marking whole passages in them, page after page, reading them again and again, even adding them to my library of micro-cassettes, so that I could listen to them on a bus, or a country walk. They became dog-eared, and 'The Shaking of the Foundations', (appropriately enough, in a way) finally fell to pieces, and had to be held together after that with rubber bands...

Was it all over then? - this herculean struggle for peace of mind and spirit, and the beginning of new life and new being? According to Gabriel Marcel, 'It is never a simple return to the *status quo*, a simple return to our being, it is that and much more ... an undreamed-of promotion, a transfiguration.' Was this it, then?

There was always that possibility, I suppose. But, with Paul Tillich, I was to discover that 'the new ... must break the power of the old, not only in reality, but in our memory; and one is not possible without the other.' From then on there were to be increasingly long passages through calm waters; and never again was the patch of blue in the storm sky to appear quite so small, or the blackness under the trees quite so impenetrable. But memory, with its sombre brood - of pain, and never-ending sense of loss, of nostalgia, ever mingled with regret - was to remain for long a stumbling block.

There are times, too - and perhaps always will be - when my 'courage to be', no match for Mary's anyway, forsakes me still - with

Each day
a lifetime:

each waking
a birth,
with its fierce pang
for the lost womb
of the night,
its first breath
deep-drawn in protest
at the burgeoning of day;

each morning
a childhood,
deprived of its
innocent obliviousness
of Man's mortality;

high noon
a middle age
of unfulfilled
intentions
failing to survive
the close scrutiny
of day;

each afternoon
a retirement, forfeit
to a frantic foraging
for the mislaid meaning
of the past and hope
with which to face
a boding future;

each evening
an old age
graced,
surprisingly enough,
by a brief acceptance
of things as they are;

and night
a return
to the darkness
of unknowing.

Earlier in these pages, there was mention of a little talk I had heard on the radio at Christmastide, several years prevoously. Entitled 'A Wink of Heaven', it seemed to me to embody the very heart of the Christian message.

Having spoken of how uncertain life can be and more often than not is, in the big issues as well as the small, the speaker went on to ask, 'What has all this talk of uncertainty to do with Christmas Day?' He continued:

> "The birth of Christ was foretold centuries earlier, and Christians say it was divinely planned before the beginning of time. And yet, when it happened, it could not have been more unrehearsed: he was born into a makeshift set-up amid people who were simply improvising as they went along.... If God was giving Himself to humanity in this child, then he was surrendering himself to uncertainty; and this is the 'wink of heaven', this is the 'whisper' that Christmas brings us - that God, in his relationship with this world, has always surrendered Himself to uncertainty. 'Emmanuel' - 'God with us' - that is the name the prophet gave to the child who was to be born. 'God with us' in all the accidental chanciness and the moral uncertainties of this life. He won't give you a nicely-painted signpost... He will give you Himself. God doesn't make things happen - he takes what has happened and he says 'What shall we make of this?' "

Yes – 'God with us' in the accidental chanciness of the hurricane and the flood, the heart attack and the stroke. He doesn't make any of these things happen, but he is there, with us, sharing the uncertainty and the agony - surely 'the one thing needful' for us to know.

Chapter 25

I am staring – no, not at a blank sheet of paper, but at its modern equivalent, a blank LCD screen. And, like the blank sheet of paper, it is inviting me to fill it with words, aided and abetted in this by a cursor impatiently prompting me, second by second, to get started. Suddenly then, I saw the blank screen as providing a strong analogy for the situation in which I found myself in the summer of 1989, a year after Mary had died.

Yes – it was as I emerged from the immediate and most traumatic effects of bereavement that life itself began to assume the appearance of a blank page, with the need to give serious thought as to how, and with what, I should fill it. At the age of 75, it was a strange, and almost daunting experience. For twenty years my life had, so to say, revolved round Mary's, despite the fact that for half of them I was still engaged in the final stages of my career. It speaks volumes for Mary's un-demanding nature that such a thing was possible at all.

To change the analogy: whatever was to help fill the huge vacuum created by Mary's death, a *sine qua non* would inevitably be its *creative* nature. I am well aware that some would regard this as mere self-expression, even self-indulgence – indeed, even as self-*ish*. Why not just settle for helping out regularly in the local charity shop, or night shelter, or whatever? Yes – why not? Haven't I already engaged myself in this debate on innumerable occasions in the past? Taken to its logical conclusion, of course, it would deem *all* art – indeed, *everything* that did not have an obvious and immediate practical application as a waste of time. So, with painting already a part of my life, albeit of necessity a very small part, why not spend significantly more time on that, now? I had a very strong feeling that deeply creative activity would go a long way towards enabling me to 'drink the full draught of unassimilated pain', and the 'unshed, countless tears'.

So it was that I joined a painting class – no, not any old painting class, but a *portrait* class. What was so special about

Plate 16 Pastel Portrait of French Student

that? Quite simply that it would involve me, in a very special way, with *people* - other human beings – a great need of mine at the time. What is more, I would have the privilege – and indeed, the *necessity* - of getting as close to their nature as was possible within the limitations of a classroom setting. As I am sure the reader must realize, a true, and *complete* portrait must somehow or other depict, not merely the geometrical layout of their features, but the *personality* that lies behind them: their sex, their age, and their character. Such involvement with other humans was something of which I was desperately in need, after my intense and 'rarefied' relationship with Mary over many years. It was, in fact, a slice of ordinary human life and relatedness, in the context of a creative activity – and *that,* I soon found, was very special. And for that reason I can still remember quite a number of the 'sitters', and my interaction with them.

Two of the (soft pastel) portraits I painted at that time are reproduced in Plates 16 and 17. Plate 16 is the portrait of a young French University student, no doubt seeing it as an opportunity to add to her pocket money. She was a very good subject, and, as was usual when we had completed our paintings, she came round to inspect the fruits of our labours. To my surprise and delight, she took one look at my work, and promptly gasped, 'I had *no* idea I was as beautiful as *that!'* It made my morning, and subsequently, having photographed the painting for the records, I had it properly mounted and framed under glass (necessary with soft pastels), and duly presented her with it. I have no idea of her name any longer, nor where she lives in France, but as I was typing these words, I had a vision of her, in her late 30s now, married, and with a family, saying to her children, 'And *that's* what Mummy used to look like when *she* was a teenager.' You see what I mean about a slice of ordinary human life...

The portrait in Plate 17 could hardly be more different. It is of 'Sylvester', a young black man, no doubt also trying to supplement his pocket money. Well do I remember quailing a

Plate 17 'Sylvester'

little at the thought of depicting such natural dignity, but somehow I seem to have managed it. And how gratified I have been to witness the response of those with whom I have shared the painting in the years that have followed! - a painting which celebrates the natural dignity of an ordinary human being, and, what is even more to the point, one with whom many would still not associate such a quality, simply because of the colour of his skin.

I stayed with the portrait class for something like two years, but then - several dozen portraits on - painting gave way to something which was to become much more significant in my life – writing. With so much of those twenty years spent looking after Mary still so fresh in my mind, I had a sudden urge to commit some account of them to paper before they had become simply part of the hazy past. So it was that I started work on a full-length book, dedicated to Mary's memory, and called 'Your Sort of Courage' (the title coming from a poem written in 1968, in the aftermath of her first stroke):

> ....The mountaineer
> in all his strength,
> roped
> to stalwart colleagues,
> with the blue sky
> and the towering peaks
> above
> to challenge him,
> knows nothing
> of your sort of courage;
> ....as,
> scorning help,
> with walking frame
> and caliper
> you move,
> step by step,
> precariously,
> alone in your weakness,
> towards
> the fireside chair.

Your Sort of Courage was published in 1994, with a first edition of 1000 copies. I believe that the publishers had some sort of understanding with voluntary organizations concerned with such matters as bereavement - that they would acquire copies for their local libraries. In the event, however, they were unable to do so (the depression of the early nineties was still making its presence felt) and the sales fell short of expectations, despite the many enthusiastic reviews.

The positive side of all this was that from time to time I was able to continue to order extra copies to meet my own needs, without any fear that supplies would run out. Imagine then, my sense of disbelief when, about ten years on from publication, I ordered another half-dozen copies, to be told that the publisher could no longer supply them. I rang them up. 'There must be some mistake,' I said, 'you must have several hundred copies still.' 'We don't, you know - not any longer.' 'He's hedging,' I thought. Smelling a rat, to put it vulgarly, I protested, 'What on earth are you talking about?' He came clean then, his tone apprehensive, to say the least of it. 'You see, we – we *pulped* them.' 'You *what?*' I was almost shouting now. 'Yes – we've *pulped* them. Didn't we tell you?' 'You're asking *me*. You know full well you didn't.'

I don't think I have felt more angry in my whole life than I did in that moment. 'I just can't believe what I am hearing. It is utterly *outrageous.*' ' You see, we couldn't afford the storage space any longer.' 'Then why wasn't I given the chance to take them off your hands? Why? *Why?* **Why**?' I was almost *screaming* with anger by now. 'We're very sorry. We must have overlooked it.' 'Overlooked *me,* you mean - knowing full well that, putting aside their monetary value, which, at this stage is not very much, they were in many other ways virtually *priceless* to me.'

By then I was feeling as though they had pulped not just the books but Mary and myself. 'Your behaviour leaves me *speechless.*' And with that, I slammed the phone down, and have had nothing more to do with the firm since. The mystery is that they were (and presumably still are) a very well-known and highly-regarded publisher, especially of what I will risk referring to as 'religious' texts, and those concerning the deeper aspects of human life and experience. Sad, isn't it? – to say the least of it.

They had probably broken their contract, but the whole episode had wounded my spirit so deeply that, even if I could have afforded it, I would not have been able to face all the hard and bitter words that would have besmirched the proceedings, had I taken them to court. Instead (as you might well have guessed) I did my best to calm the emotional storm that continued to rage inside me (make no mistake – I was almost *beside* myself) by writing a poem about it – and a very caustic one at that:

Reduced to a Pulp

Had I but known,
I could have intervened –
and saved at least a few
(among several hundreds)
from a violent end.

None of them a stranger
to my deepest thoughts, all
privy to my greatest sorrows,
each, a witness to my direst
pain – they were almost
part of me; to renew
the acquaintance of
but one of them a source
of strength, should courage
wane.

Those who bear the burden
of their untimely end,
though – supposedly –
custodians, were yet
prepared to sacrifice
their protégés, for a
trivial monetary gain –
wholly disregardful

of the investment
I had made in them:
not of money, but of
my very self.

Hard indeed, is it, to
believe that the hundreds of
copies remaining, of a book
in which I had bared my soul,
shared the worst and best
moments of my life, had,
with not a word of warning,
been summarily *pulped*.

I, too, with them –
or so it seemed.

No - I did *not* send the poem to the publishers. If they hadn't the sensitivity to realize what pulping several hundred copies of the book would do to me, they certainly wouldn't have what it takes to respond to a poem about it.

Incidentally, acting on a sudden impulse, I have just typed into 'Google' the words 'Your Sort of Courage', and to my great surprise, there are still both new and used copies available from the likes of Amazon Books. Indeed, imagine my delight to find a so-called 'Editorial Review' on the Amazon Website, which was both detailed, and very appreciative – and *that,* fifteen years after publication... So, here it is:

" 'It was 7.30 on the morning of Monday June 17th, 1968, and even at that early hour the weather augured well... Mary had as usual followed me to the door, and stood watching and waving as I bounded up the steps of the little snicket which led to the next road, and my shortest route to the station. I looked back briefly and waved, and as I did so she turned into the house again - and closed the door on an era.'

"When Lloyd Kemp was next to see his wife, she was in a comatose state, having suffered a massive stroke. The hopes for a new lease of life together, with children grown-up and retirement ahead, were dashed with one brutal blow. This is an account of a couples' struggle to maintain a loving and fulfilling relationship while struggling against the physical and mental effects of the stroke's aftermath. Partly biographical, partly autobiographical, the book throws light on Mary's ability to inspire all those whose lives she touched. At the same time it gives an honest account of the author's search for understanding of the nature of personal suffering and the spiritual problems raised. The book is peppered with Lloyd Kemp's heart-felt poems, reflecting his life with Mary, and through these we gain a clear picture of both the author and his wife - two ordinary people whose circumstances compel them to call on extraordinary inner resources of courage and strength. This is a story which is bound to affect the lives of many others. Carers and the cared-for alike will find solace and comfort in this account of courage in adversity, and will recognise many of the sentiments expressed throughout this work. "

And there was also a snippet which I picked up from the website of The Royal Society for the Promotion of Health, where one Jane Watkins speaks of the book as being 'captivating and intensely moving'.
So, I take heart – 'Your Sort of Courage' lives on, despite the publisher's best efforts to put an end to it.

Writing Your Sort of Courage did much more than filling the vast vacuum left in my life by Mary's death: it was, as I foresaw, enormously *cathartic,* going a very long way towards enabling me to plumb the depths of the experiences we had been through together, on our journey through those twenty years.

We have reached 1994, the year in which Your Sort of Courage was published, but also the year in which I celebrated my 80th birthday, and which I celebrated in Germany with Rosemary and her friends. Following the fraught visit at Christmastime 1988, I had continued to visit her each year, though never again at Christmas. These regular visits provided me with the necessary impetus to learn some German, which (despite the fact that I would in no way rate myself as any sort of linguist) I did quite successfully, with the aid of the excellent software by then available for almost every language on Earth. I worked quite hard at it, and, thereby, making my visits to

Germany all the more enjoyable. And with such visits stretching out ahead for maybe another ten years or so, I certainly regarded the time as well spent. How was I to know that in the Spring of 1996, the visits would be brought to a sudden end – an end which I couldn't possibly have anticipated?

When I began writing Your Sort of Courage, I soon realized that there was no way in which I was going to be able to run the writing and the portrait painting in harness: I would be revisiting - and reliving – the most harrowing moments in my life, and the writing would need, not only my undivided attention, but also all the creative energies I could bring to bear on it, if it was to be brought to fruition. The outcome was not only (as I have said) deeply cathartic, but, as we have just seen, it was actually well received. Inevitably, it brought to mind that earlier episode, when my short story, Two Brass Vases, had been deemed worth purloining, to appear under another author's name in a well-known magazine. Putting the two things together, I found myself wondering whether they indicated that I had some sort of talent for writing which I should explore further: why not try my hand at fiction?

So it was that by 1996 I was deeply engaged in writing eight 'long short stories', as the genre was designated, each of them several thousand words in length. They did have, in fact, a common theme - *surprise encounters* - with only of them in any way autobiographical: it was entitled Marion's Field, and featured the painting reproduced in Plate 18.

I have just read it again for the first time for some ten years or more, having, during that time, forgotten all but the general gist of it: its mystical nature, and the surprise ending, completely lost in the mists of time. A major part of the story is strictly autobiographical, recounting how one Henry Atkins, 'elderly but still very active', had walked away from his wife Marion's grave, wondering 'whether he would ever again have the incentive to continue with even one of his many interests', and of how, only five days later, and despite such feelings, he had painted 'Marion's Field'. The story continues - telling how,

Plate 18 'Mary's Landscape'

the following Spring (as *I* had done), he had returned to produce a second painting, depicting all the changes which springtime had brought about in it.

At this point the story becomes fictional, and highly symbolic. As he paints, he becomes aware of a young woman, beautifully dressed in a Spring outfit, tripping towards him from the far side of the field. She takes an immediate, and keen interest in the painting, and as they share his coffee and biscuits, she suddenly, and dramatically, asks him if he will paint her portrait. Her question startles him, since she seems to be privy to the fact that he had recently been drawn to portraiture, and was attending classes. Despite that, he is by no means keen to 'rush in, where angels fear to tread': for him, there is an aura of mystery about her, which, somehow, would have to be incorporated in any painting of her, and he is far from convinced that he could succeed in such a task.

Despite this, he finds himself unable to turn down her invitation, drawn by the very enigma of her being, and makes an arrangement for a first session in his makeshift studio the very next morning. Despite his misgivings, the two-hour session goes very well, and, thrilled with the results, she is only too ready and willing to attend a second, and final session, the morning after. Then:

> "Suddenly, she realized that he was no longer working, and appeared instead to be agonizing over the painting. She got up at once and went to him.
>
> 'What's happened?' There was deep concern in her voice.
>
> 'It is finished,' he murmured, almost inaudibly.
>
> 'No! No!' she whispered, as she put her head close to his. 'It's just the end of a new beginning.'
>
> For the moment his vision had forsaken him, and he could not follow her into the future which she was plainly foreseeing for him. He clung to her desperately, as she studied the painting, which had brought her to the point of tears.
>
> 'I will have it framed for you - I can get it done this afternoon, and you could come for it tomorrow.' He was playing for time - anything

that would bring her back again - convinced nonetheless that they were moving inexorably towards a final parting.

'The painting is for you, not me,' she was saying, then. He looked up, incredulous, convinced that he must have misheard her.

'It is for *you*,' she whispered again. 'I always intended it so.'

His problems over her identity - no, the very nature of her being, came to a head then, and he clung even more desperately to her, as though afraid now that she would vanish before his eyes.

After a while she gently freed herself, and quietly left the room. A minute or two later, and he heard the outer door close softly behind her.

He sat, transfixed by the course the morning's events had taken. Then, slowly, resolutely, he stood, and went to the sitting room, where he retrieved the newly framed and glazed second painting of 'Marion's Field'. With a razor blade he carefully sliced through the fresh brown paper tape sealing the back of it, removed the retaining tacks, and finally the painting itself.

He took the portrait from the easel, arranged it in the empty frame, and propped it up on the chair where she had been sitting just minutes before. Then he put the painting of the field on the easel, and began to work on it, repeatedly looking across at the portrait for inspiration.

Slowly and deliberately, and with consummate skill, he painted her figure in, just as he had seen her for the first time, walking towards him along the field edge, her golden skirt dancing about her, hair streaming in the breeze, her feet seeming barely to touch the ground.

When he had finished, he leaned back on his stool and surveyed his handiwork. And in that moment it came to him that the painting of 'Marion's Field' was now complete.'

It was a dark evening in the Spring of 1996, and I had been working hard on 'Marion's Field', and was ready for a break. I consulted the weather, only to find that it was drizzling with rain. I had been contemplating walking the mile along Englishcombe Lane to my friend Joan, who, not long before Mary had died, had lost her husband, Bill. We had similar tastes in TV drama and sit-coms, so once or twice a week we kept each other company as we watched the latest episode of one of our favoured sit-coms, or whatever. 'Drat the drizzle,' I thought, as I began to settle to some more word-smithing. But I had had enough of it for one day, and I was soon donning my raincoat, to set off for an evening with Joan. It was a fateful decision: one of those that one lives over and over again, as I might well have read one of my stories over and over, with the prospect of being able to change its outcome. But real-life decisions aren't like that, are they? – there's no back-tracking.

I was walking in the pitch-dark, and parts of the pavement had been recently patched up, and badly, too. So I took to the gutter – it seemed a reasonable enough thing to do, as I would be facing the oncoming traffic. All went well – that is, until one of the approaching cars I deemed to be travelling far too fast, and too near the curb for my liking. Of course, he could see me in his headlights, and in fact there was no danger that he would knock me down, but I didn't fancy the close encounter, and made to get back onto the path again as he came towards me. Disastrously, I chose to do this by starting to take a step onto the pavement with the foot that was next to it. It caught the curb and, with the other foot on the other side of me, and thus unavailable to save my fall, I went over like a felled tree, and as helpless, the head of my right femur hitting the pavement like a hammer. I remember thinking that my body had *never ever* suffered a blow like it. Strangely, at the time, it didn't actually hurt very much, and I went to stand again, but the leg promptly collapsed on me, and I was on the ground again, and hurting badly this time. Although I didn't know it, my femur had snapped, just below the head. I managed to flag down another oncoming motorist, who went into a nearby house to phone for an ambulance. So it was that, instead of arriving at my friend Joan's for a happy little get-together, I found myself on the way to the hospital.

After scrutinizing the X-ray, the surgeon said, 'Well, if you were *determined* to break you femur, you've broken it in just the right place.' Two days later, and I was in the recovery room, with the two parts of the bone now held together by a stout metal strip of some sort or other, which would remain in place - part of me - for the rest of my life.

Remember - I was 82, and such an accident was a very serious one for someone of that age, and I was nearly a month in hospital, before being allowed home again. Even so, there were inevitable difficulties arising from living on my own, one of the most trying being that I dare not venture down the two steps into the garden by myself. Well do I remember sitting just inside the door, feeling like a prisoner in my own house, the garden so near, and yet so far. And there were so many – even little – things, which for the time being became difficult, or even impossible to do: like making a cup of tea, and then, instead of carrying it to the dining table, having to slide it several feet along the worktop to the point nearest to the table, before sitting down and reaching out for it. And so on, and so on.

The fall had been in March, but many of the difficulties were still there when summer arrived, and, almost inevitably, I fell victim to 'post-traumatic stress disorder'. My doctor put me on an antidepressant, but it didn't seem to help much; then, suddenly, she suggested that I should go to stay with Rosemary for a week or two: I needed a change, she said, and it might do the trick. It didn't. Sleeping (or trying to) in a strange bed, and soon insomnia had been added to my problems. This was compounded by the serious difficulties I was still having with my walking: remember, it was a mere three months after breaking what is the largest bone in the body. Moreover, Rosemary lived up a steep hill, and a visit to the town centre inevitably involved climbing the hill on the way back. Another problem – all too familiar to me – was that (it being the middle of the Summer term for her) I was spending too much time on my own, and away from the resources of my own home (such as my computer) with which I could have occupied myself.

And the consequence? – I brought my stay to an early end and returned home, with the insomnia as extra baggage. My friend Helen met me with her car at Heathrow, and our first port of call on reaching Bath was not my house, but the doctor's surgery. I wasn't able to see my own doctor, but, instead, saw a colleague of hers - also a lady doctor. She listened to my tale of woe. I was close to nervous exhaustion, owing to the lack of sleep, she said, and though it would mean Temazipam for a little while, it had to be. It was all too reminiscent of the run-up to Christmas, 1988: suddenly I was no longer on a ladder, but a very long snake threatening to take me back to Square One again.

However, in the event, and blessed by a deeply-caring doctor, I soon made up the lost ground, to find myself back in harness again, and putting the finishing touches to those eight 'long short stories'. For myself, I have to say I was really pleased with them, but I soon found out that that was not the case with the two or three agents to whom I submitted them. I've forgotten the 'in' word they were using: they were not 'experimental' enough, or some such. Typically, I hadn't done any market research before embarking on them. I had simply written about situations (and people's response to them) which I felt typified the human condition: I would have felt it a waste of my time and of the wonderful writing facilities provided by a word processor, to write in such a deliberately enigmatic and fanciful manner as to have the reader wondering, 'And what was all *that* about?'

It might be asked, 'Why put pen to paper (or chase the cursor across a computer screen, leaving a trail of words behind) – why go to all that trouble, with rather more than an inkling that the words were never going to see the light of day streaming through a publisher's office window?' A good question... And the answer? I suppose it has something to do with the futility of trying to stop the jet of water from an artesian well by sitting on it: such is the creative urge - at any rate in my experience...

The bug had bitten deep, and - to indulge in a dreadful mix of metaphors - it was seducing me with the prospect that if I embarked on what would be a *far* bigger undertaking than the eight long short stories, I could kill two birds with one stone: what about writing a full-length novel now, about a young lad who, from a very early age, shows enormous promise as a classical composer - a writer of great symphonies? That way I would, in fact, not only be satisfying the ever-present urge to create, but also, in my imagination at any rate, I would living the life that I would have had no hesitation in choosing, had I been offered such a choice, even before my age had reached double figures. Moreover, writing such a novel would - through the mouths of its characters - provide me with a natural setting for all those things I had long been wanting to say about the role of great music in the lives of us humans.

    I had recently read a little book on 'writer's block' – out of interest rather than need at that moment (having, of course, just finished writing the long short stories). It advocated a simple, but strict routine for the writer: in a nutshell, 'rise early, and get on with it'! The theory was that you are not only *physically* fresh after a night's sleep, but that the *mind* - not yet cluttered by the business of a new day - was at its freest, and most creative. It suited me fine: I had always preferred being up with the lark, to sharing the small hours with the owl. And a new routine to the day seemed a fitting setting for the new project.

    I made a start in 1997. Often titles have to wait a while to emerge, but not in this case: right from the start it was to be 'Michael's Notes', not the kind that you write in your diary, but musical ones. In the extracts which follow, 'the squire' is the latest incumbent at the local Manor, where Michael's father is the general factotum, the squire being Michael's self-appointed patron from the moment he had learned of Michael's budding genius. Mary is, of course, Michael's girl friend. They are in their middle teens, and the occasion is the greatest in Michael's young life – the premiere of his first Symphony.

'Huge forces were at work that night, and were beginning to combine. There was the power of music - not just of Michael's music, but of music itself. And not music in the abstract, either, or music disembodied by having been transferred to a gramophone record. It was *live* music - music to which men and women were giving birth there and then, and the struggle and effort and joy involved were there for all to witness.

'It was the power, too, of joint human effort at its most skilful - its almost hypnotic power, as the orchestra, well-nigh a hundred strong, put Michael's notes together in perfect tune and time.

'And it was the power of human beings, gathered for a common purpose - to respond more deeply and more completely in each other's company, than they would, or could have done, individually.

'It didn't end there. For Michael's family and friends, all together in the squire's box that night, it wasn't just the sense of occasion, the sense of being part of that vast audience - of that vast sea of faces below them. It wasn't simply the spell cast on them by the sight and sound of an orchestra playing in perfect ensemble. It wasn't even the power of music as such. It was the fact that it was *Michael's* music - that every one of the myriad of notes that made up the concert that night had come into being in his imagination, before it could ever be heard by any of them.

'There was a buzz of expectancy everywhere, heightened still more as the oboe began sounding an "A", and the rest of the orchestra joined in, with seemingly frenetic enthusiasm. The squire and Michael looked at each other, sharing another of those moments of which they would never tire - when the necessary discord of instruments being coaxed into tune held such limitless promise - on this occasion nothing less than the Symphony to come.

'[The conductor] called them to order as the sound died down, and cast a gently enquiring look in the direction of the principal flautist. With lips pursed and flute aloft, she nodded her readiness. A moment later, and he had launched her on her celebration of young love - a shepherd's pipe on a distant hilltop, with a descant of violins soaring high above. The flute faded, and its sound was replaced by the mystery of muted cellos........'

And so on, and so on. I had the bit between my teeth, and for two years I followed the advice of the little book, getting up early, flinging on a dressing gown – not bothering to dress – and writing several hundred words before breakfast: I can thoroughly recommend that little book's advice.

Towards the end of 1999 Michael's Notes was finished – all 210,000 words of it. During the following two years I sent off a synopsis to two or three literary agents. One of them showed some interest in a general sort of way, but encouraged me to try other agents, who might be more equipped to handle a novel on classical music: it wasn't - I suppose one might say - 'up their street'. Or, more likely, were they just politely 'fobbing me off'? To sum it all up, my long short stories were too 'straightforward', and the subject of my novel was of interest to only a small section of the public. And the upshot of that? - that I just couldn't bring myself to spend the time and energy that would be required to 'go the rounds', touting my wares, so to say, only to find eventually that I had been flogging a dead horse. Mixed metaphors again! – but let's face it: what chance had I with a novel on classical music, when the first Harry Potter novel was turned down *twenty times* before it got its foot inside a publisher's office. I had in any case achieved my basic objective, which was to put on record (albeit, it would seem, just for my family and my friends) what music meant to me personally, and, indeed, what I saw to be its crucial role in this human life of ours.

Chapter 26

Despite all that has been said above concerning three major *writing* projects that occupied most of the 1990s, the reader must be left in no doubt that the composition of *music* has nevertheless been like a golden thread woven into the very fabric of my life – ever since the time when, well under ten years of age, I was found drawing lines, and putting little black blobs on them, solemnly declaring, when quizzed, that I was 'writing music'. And let there be no mistake, there *is* a visual beauty in a music score, before it is ever converted into the sounds it represents: there are those who have framed facsimiles of part of the score of a great composition, and hung them on their walls. And well do I remember visiting a museum in Vienna, where (under glass, of course) the original score of one of Beethoven's symphonies was on display: there seemed to be a positive *aura* about it, and I could have wept with 'holy joy', as I gazed on it.

So - perhaps it has been rather remiss of me not to have drawn attention from time to time to the on-going presence in my life, of that golden thread. To make some amends now (as they occur to me), there was the carol that Mary and I wrote for the first Christmas of our married life – Mary writing the words, which I then set to music. Of course, expecting our own first child John (the following May/June) made that Christmas very special for us anyway. Many years later (in 2001), I arranged the carol for four-part choir. This was, in fact, a little Christmas present for Bruce, who has a rich bass voice, and is an enthusiastic member of his church choir in Perth. Indeed, reciprocating the gift, he subsequently presented me with a DVD of its premiere. As well, in 2004, I wrote the words and music of a little Vesper for the selfsame choir, which gave it its first (and only) performance – but of which, once again, I have a DVD.

Then there were the middle 1960s, when our family trio enjoyed an all-too-brief span of life, before the children – fast growing up – went their separate ways: Rosemary, of course, the flautist, Roger, the violinist, and little old me on my cello. I wrote several Trios on which to flex our musical muscles, but only one has survived. It is in classical form, and in its way quite sophisticated – but well do I remember my Aunt Cis, whom we had invited for a little stay, paying it a doubtful compliment after we had, with great gusto, performed it for her. 'Pretty little tunes,' she pronounced...

By then a competent performer on classical guitar, John was in a position to make his own particular contribution to family music-making, and incidentally provided another outlet for my insatiable urge to compose. I wrote quite a number of pieces for him, some of them – notably 'The Troubadour's Song', 'Soliloquy', and 'Courante', being published in the BMG (banjo, mandolin, and guitar) Magazine.

Strangely (or is it so strange?), writing music lapsed almost entirely during the twenty years in which I was so deeply involved in caring for Mary, and - as we have just seen – also for the ten years following her death. That's not *quite* true: during the early 1990s I 'kept my eye in' (or should I have said 'my ear'?), by writing a whole series (48 in all) of what, in Bach's time, would have been called 'two-part inventions'. These were specifically written for my German grandson Kevin, who had started to learn the piano, and in due course I presented him (a rather extravagant gesture) with a leather-bound copy of them. Not long after, he gave up the piano! (but I try not to feel I had any part in that...)

And then there was that remarkable episode in the Autumn of 1996, the year in which I had broken my femur. I was still very depressed, but, despite that, had written a short piano piece called 'Joie de Vivre'. Surprised? *I* wasn't: it is well known that some of the happiest music has been written by composers whose lives were far from happy at the time.

It was in the evening, about 9 o'clock, and I had just played my new piece through with almost gay abandon. Acting on an impulse then, I switched on the television, but with no intention of ending my session at the piano: just a little break from it (or so I thought). I was immediately *galvanized* by the programme (a documentary) which had only just started. I've forgotten its title, but they were giving a brief outline of its gist. Jung would have said that it was a prime example of what he termed 'synchronicity' – something much deeper than mere coincidence. (As Wikipedia puts it: 'Synchronicity is the experience of two or more events which are *causally* unrelated occurring together in a *meaningful* manner. In order to count as synchronicity, the events should be *unlikely* to occur together by chance.')

We are into strange waters. You see, just as I switched on, still exhilarated by the feeling of having created an effective little piece of music, they were saying that the programme was to concern itself with *the therapeutic role that creativity could play in overcoming depression.* I could hardly believe my ears, and, even as I recall the whole incident, I find I am still coming to terms with it. The sudden impulse (to switch on the television at that particular moment) felt *purposeful,* not impulsive, and the outcome was to find myself presented with the results of a serious piece of scientific research confirming *what I had, only moments before, discovered for myself* - that allowing oneself to be taken over by the creative spirit had, at any rate for the time being, lifted me out of my depressed state. 'That may well be true,' I hear you say, 'but after all, the piece of music you created *was* called "Joie de Vivre" '. That's true, but I am convinced that had it, instead, been called 'Soliloquy', or 'Nocturne', or whatever, the results would have been the same: it is *the act of creation* itself which is therapeutic, and, I believe, is not dependent on the nature – or the excellence or otherwise – of what is created.

In these contentions, I am backed up by the work described in the TV programme. A group of some half dozen or so of seriously depressed adults of both sexes, had been provided with a whole series of creative tasks in which to engage, over a period of some months. The programme began with shots of members of the group before the research began

– their general behaviour, and their conversations with one another - leaving the viewer in no doubt whatever of their depressed state. And it ended with them punting on the Cam, along past the grassy Cambridge college 'backs', the scene positively 'oozing' serenity!

I have dealt at some length with what happened on that September evening way back in 1996 for two reasons: it was, in my opinion, a startling example of the phenomenon of 'synchronicity', whilst at the same time providing firm evidence that creativity is not concerned simply with producing works of art, or inventing clever devices, but has a role to play in the mental well-being of all of us, even in our everyday lives.

Now - writing out music (as distinct from composing it) can be an awful chore (despite my childhood addiction to it!), but there was no equivalent of a word processor (or so I thought) to eliminate the tediousness of producing a fair copy of what one has jotted down with the aid of a friendly pencil (and the facility it provides for easy erasure of mistakes...).

'No equivalent of the word processor for music', did I say? It was just as I was putting the finishing touches to Michael's Notes in 1998 that my friend Dick chose to introduce me to a piece of German software called 'Capella'. It was immediately obvious that I hadn't been moving with the times. It is true that I'd heard of the existence of software used by professional composers to help with the exacting and tedious work of orchestration. But it was beyond my wildest dreams that modestly-priced software was already available, which would enable amateur composers the likes of myself to produce computer files of their compositions with ease, from which copies could be printed, looking for all the world as though they had just left the printing press of an established music publisher.

Of such ilk was 'Capella', and I allowed very little grass to grow under my feet before I was installing a copy of it on my computer. It was very well written, and it wasn't long before I had mastered it, and was producing hard copies of my favourite compositions. Without anticipating the story behind the particular composition reproduced in Plate 19, I would draw the reader's attention to it as an example of the sheer excellence of the end-product produced by Capella.

Capella unlocked a door in my musical life that I had come to think would always remain closed: I had long since concluded that it was, and would remain, a very private affair, since the only means I had of sharing my music with others (or so I thought) was via Xerox copies of my hand-written manuscripts, which I felt was expecting a little too much of any would-be performer thereof. But, almost overnight, Capella had changed all that: *now* I would be able to present anyone interested with a copy of *any* of my efforts at composition, which would be indistinguishable from any other sheet music. It had added a new dimension to my musical activities, and, that being so, it was not surprising to find that, after nearly ten years of word-smithing, the urge to create - in full flow again now – became re-directed *in toto,* to composition.

There was a song cycle ('Beyond the Darkness') for tenor and piano accompaniment – settings of five of the poems I had written for, or about, Mary; a cantata ('Now can I see') for four-part choir and tenor solo, with string quartet (or organ) accompaniment (a setting of a long, narrative poem I had written years before, and based on the story of the healing of the blind man - John, Chap. 9); and a whole series of Piano Suites, together with a Sonatina for Piano. Dick's friend Antony, a professional pianist, recorded a number of the piano pieces - together with the song cycle (with a tenor friend, Peter Uncles), and presented me with the whole collection, put together on a single CD. I was able to make copies of the CD, label them quite professionally with the aid of dedicated software, and present them in the usual 'jewel cases', provided with the appropriate insert and booklet giving technical notes, together with the words of the song cycle. The end-product was such that it could have graced the shelves of any CD store!

Plate 19 'Lament for Mankind'

And was I proud of it! – and, yes, *moved* by it: well do I remember quietly shedding tears as I listened for the very first time to the song cycle I had written in memory of Mary.

But I must return to the piece of music depicted in Plate 19. I had got back home again about 2 pm one afternoon after shopping and lunch out, and had switched on the TV before briefly dropping into a chair to rest my weary legs. And as the screen lit up there was a shot of the New York skyline, with a huge cloud of smoke and dust, backed up by flame, emerging from half-way down one of the skyscrapers. Mere moments after, and to my disbelief I saw an airliner fly straight into a second, and similar, building alongside the first one. I was just beginning to think what clever trickery modern film makers get up to, when I became aware that the voice over was giving a *live* commentary, and what I was seeing on the screen was at that very moment *happening* - 'before my very eyes'.

It was, of course, '9 11', 2001, and I had just witnessed hundreds, if not thousands, of human beings going to their deaths, in a split second. I sat numbed – petrified – by what I had seen, and began to feel an overwhelming need for *some*-thing, which, some-*how,* would help me even *begin* to face the horror of what was taking place. And it was music, of course, that came to my rescue - and the opening bars of 'Lament for Mankind' came into being. By the end of the next morning it had been completed (Plate 19).

What better example could one have of the role that music can play – not just ceremonially, but in a vitally practical way, in our lives?

Those three years, from the end of 1999 to the Autumn of 2002, were halcyon years for me. I continued with the same morning routine that I had adopted when I was writing 'Michael's Notes' and the short stories, but there was one notable difference. To begin the day by sitting down in front of the computer and reading, sentence by sentence, the words one has written the previous night is inevitably something of a chore. But, by comparison, to begin the day by sitting down at the *piano,* and *playing* what one has written the night before – *music,* not words – well! – that's a very different matter. As I recall such moments, even after ten years, I find myself re-living their 'Eternal Now-ness'. It's nothing to do with *my* music: it is music *in general,* with its ability to lift the spirit almost instantly into a milieu which is above all life's pettiness. And what a blessing *that* is, first thing in the morning.

Writing music was virtually a continuous process during those three years - there being, almost always, a piece 'on the go' to wake up to, and to serve as my 'Open, Sesame' to another day of 'holy carefreeness', as I chose to describe it, in a recent poem. Of course, one couldn't expect to stay on such dizzy heights for long: there were three meals a day to prepare, eat, and clear away, to say nothing of the regular shopping expeditions involved. But somehow, with the latest composition forming the backdrop to the whole day (and thus never far away), time itself seemed to become – what shall I say? - *musical* time, characterized, not by transitoriness, but by the 'rhyme and rhythm' of Creation itself.

A bit 'over the top', you say? Maybe. But the experience of being involved with actually creating music does have a feeling of timelessness about it impossible to describe in words, and in my experience approached only in the peak moments of writing a poem. If that is true of one of *my* humble creations. then I have to ask myself, 'What must *Beethoven* have felt like immediately after putting the finishing touches to the Ninth Symphony?' I would not dare even to attempt putting it into words: I am filled with awe at the very thought of it. And I hasten to add that my choice of 'awe' is in no way an attempt to impart a spiritual or religious 'flavour' to the point I am making: of all the words available, 'awe' seemed to be the one that got anywhere near to the feelings I experienced as I tried to put myself into Beethoven's shoes. Such thoughts and feelings need no 'explanation' by any school of psychology, philosophy, or religious belief - they are what I call 'self-validating', their relevance to the ultimate questions Mankind asks about himself, and what it means to be 'human', being instinctively recognized, and immediately apparent.

Of course such 'timeless' experiences are not confined to music - but Mankind's use of, and response to, music most assuredly give it a unique place in his life. To begin with, it can be said that any formal occasion, solemn or joyful, is barely imaginable without it – from weddings and funerals, to almost all public, and state occasions. And it is highly significant to realize that if such events are televised world-wide, the *words* associated with the ceremony would need to be translated, for them to be understood and appreciated; *but not so the music.* The local music culture (e.g. Indian, or Japanese) may differ significantly from others, yet even modest exposure to another music culture will result in a significant degree of understanding and appreciation of it. Not so with the multiplicity of languages, where proliferation results in insularity that is all too evident.

One final point on this matter of the 'naturalness' of music: I am very conscious of the fact that my lifelong interest and involvement in composition has led to my having acquired a considerable knowledge of harmony and counterpoint, and musical form and structure. But I am *equally* conscious that when I am listening to music with someone *quite devoid* of such knowledge, their involvement with the music, and their response at the deepest levels *to* it, can and does, in all respects, match mine. It is almost as though someone who has made no study of German at all, listening to one of Rilke's poems, was able to understand and appreciate it every bit as much as someone who had studied German to A-level, or whatever. Music is 'in our bones', so to say. No! - it is in our very souls.

So, you can see, now, why those three years from the autumn of 1999 to the autumn of 2002 - spent virtually steeped in music - seem to have been the very acme of my creative life. But you can have no inkling of what brought that time to such a sudden and traumatic end – literally overnight.

Chapter 27

There came the day when - still occupying Cloud Nine, with my music - I suffered a deeply traumatic emotional experience. Whilst not being entirely responsible for what happened then, I am convinced that it was a very significant, if not major factor in causing the catastrophic change which took place in my hearing during the course of a single night, in September 2002. (This had, in fact, been preceded in the Spring of that year, by my right ear becoming just a little deaf, and displaying some slight problems relating to the discernment of the musical pitch of notes about two octaves above middle C, but it was diagnosed at the time as nothing more serious than 'glue ear'.) I shall say no more about the traumatic experience itself, except that it was inflicted by someone I had come to regard as a close and dear friend, and that, *emotionally,* it was among the worst I had ever suffered, in my long life.

Just three or four days after the blow had fallen, I went to bed sad and sickened by what had happened, but physically, it would seem, none the worse. But when I woke, it was with a sense of now having *two* 'glue ears', and once I was up and about, the attempt to listen to the BBC News soon made me realize that I had gone quite deaf - needing to turn up the volume on the radio by a considerable amount in order to make it intelligible. That was bad enough, but little did I know that there was worse – far worse – to come. Remember, I was listening to the *News* – to the *spoken word.* But, as was my wont, having listened to the headlines and the first two or three major items, I switched over to Radio 3 to steep myself in some music before the News had got under my skin: it mattered little what the music would be – Radio 3 rarely disappoints. But it wasn't music at all – *it was cacophony.*

I took myself off to my doctor, and asked for an appointment with an ENT specialist, which duly took place. He was in no doubt: I had *two* 'glue' ears, he said, and comforted me with the assurance that it would probably 'go away of its own accord' in a matter of months. And if it didn't, he added that there were things that he could do, as he put it. I accepted my fate, having been given that assurance. But, even as a mere physicist, I ought not to have done: I could, and should have challenged his diagnosis on the spot.

You see, 'glue' ear is a condition of the so-called *middle* ear, which (in case you've overlooked it - as I had) is the cavity between the ear drum and the *inner* ear, and three linked 'ossicles' in it transmit the vibrations in the ear drum (produced by sound waves) to the membrane comprising the interface with the inner ear, which in turn transmits them to the *cochlea*, that well-nigh incredible piece of anatomical machinery which analyses complex sounds into their component frequencies ('fundamental and harmonics') and transmits them as neural signals to the brain, where the information is 're-assembled'. Thus the middle ear simply *transmits* the vibrations constituting the sound, and 'glue' ear (in which the middle ear fills with a glue-like fluid) causes *deafness:* by its very nature it cannot *distort* the sound - any such distortion being associated with a malfunction of the *cochlea,* situated in the *inner* ear.

So it was that I lived in a fool's paradise for several months, paying several visits to the specialist, and parting with around £500 pounds altogether for the privilege of doing so. Then, however, came the time when he greeted me with a thoughtful look, and, without further preamble announced that he was going to hand me over to one of his colleagues. 'In fact,' he added, 'he can see you tomorrow.' Perplexed though I was by this sudden move, I remember thinking that at least something was happening now, and duly kept the appointment.

His opening gambit was *very* revealing, and a pathetically obvious cover-up for his colleague's shortcomings. ' 'You know,' he said, his embarrassment all too obvious, 'it's not at all uncommon for us to hand on a patient to a colleague.' 'Oh,' I said. already wondering where all this was leading. 'You see, I'm more familiar with your problem than he is. It happens,' he added as casually as may be. By now I just wanted to say, 'Why don't you cut the cackle, and simply say he got it wrong, and was too embarrassed to say so to my face.' But I didn't, of course, being polite little me. 'I think I know what is wrong with your ears,' he was saying. He didn't give it a name, and I cannot after all this time remember the exact words he used to describe the condition - at which point I felt sure that 'Google' would come to my rescue. It didn't. For a whole hour I've sat, entering every possible combination of words I could think of – from 'loss of musicality in the ears' to 'mal-function of the cochlea'. But, no go - so the condition must remain nameless.

The specialist was still talking. He was saying that just as in Meniere's disease the trouble was due to a very small excess of fluid in the semicircular canals, causing giddiness, so, in my case, it was a very small excess of fluid (I think he mentioned a few cubic millimetres) in the cochlea that was causing all the trouble. (I took him to mean that those numberless fine hairs responsible for analysing sounds into their constituent frequencies were fluid-logged, and therefore unable to do their job.) 'Fortunately,' he went on, 'there's a simple treatment, which is usually effective.' At the risk of appearing to make a sick joke, it was 'music to my ears.'. But, to my dismay, what he described then was, to my way thinking, like using a sixteen-inch gun to shoot a rabbit, as they say: I would be put on a diuretic (for three months - I think he said).

So, to remove a few cubic millimetres of fluid from my inner ears, what I believe would be several pints of fluid would be removed from my body. To put it mildly, it hardly seemed the most elegant of solutions, but I was desperate. The interview I am describing was in the summer of 2003, nine months on from that catastrophic overnight onset of the condition. During that time there had been a few partial remissions for a few hours, and – very occasionally – even for one or two days. (I came across a slip of paper a little while ago on which, at Christmastime 2002, I had ecstatically scribbled the words 'Heard a carol today as *music! Wonderful!* – I mean, that it should have been a *carol!*' Indeed, I well remember taking that as some sort of omen. But, as was so often the case, disappointment followed hard on its heels.

This might be the moment to emphasise the truly catastrophic nature of what had happened to my ears. Well do I remember over-hearing some friends *and* family to whom classical music meant little or nothing saying, in effect, 'I don't know why he's taken it so *hardly*. After all, he's only gone a little hard of hearing – don't we all? - sooner or later.' I couldn't remonstrate with them – they wouldn't have understood. How could I find the words to say that *I no longer felt I belonged to the human race* (they would probably have thought I had lost my marbles as well as a bit of hearing). But I am *not,* repeat *not,* exaggerating. I am sure that I have said enough in all that has gone before to leave the reader in no doubt that for me music, in all its different aspects, is part of what it means to *be* human.

Let me give you a practical example. You would expect that, music meaning what it does to me, I would have many friends of like mind and feeling. That's true - I do. And they continued to visit me, but, sadly, we could no longer share our love of music, or listen to it together: in effect it was a *no-go* area between us. But naturally, sometimes one of them would forget. And one such occasion is seared onto my memory, simply because of the effect it had on me.

We had spent some time talking about this and that, with music definitely 'off-limits' – that is, until the time came for him to go. 'Well -I'm going to a concert tonight at the Forum,' he announced joyfully. It was as though I had been struck a vicious physical blow. He may well have realized what had happened, and regretted it immediately afterwards, but the damage had been done. As the door closed on him I literally felt I was on another planet, and the only being occupying it. Yes – it was literally true, I *did* longer feel I belonged to the human race, and at that moment the sense of isolation was well nigh unbearable.

Back, now, to my interview with the specialist. There was a slight complication associated with the diuretic treatment: I had mild hypertension, and was on 25 mg of Atenolol to treat it. Adding to my medication a diuretic (itself a treatment for hypertension) would mean my doctor organizing the phasing-out of the Atenolol, and the phasing-in of the diuretic: a bit of a complication, I was told, but quite a common procedure. So – the stage was set: my hopes and expectations running high - it was only a matter of time now, before I rejoined the human race. *What* a prospect!

As far as I can recollect, it took a few days – perhaps a week – to make the change-over from Atenolol to a diuretic, and it proved uneventful. One afternoon, probably about a week or ten days later, I had had to deal with a visitor, regarding a business matter of some complexity, and during the course of his visit I had begun to feel unwell, in an odd sort of way, and was indeed glad to see the back of him. Something (I can't think what) prompted me to take my pulse. *It was just 30 a minute...* I rang my doctor, but she was not available, so I spoke to one of her colleagues, and gave him a brief run-down of the situation: his verdict - that I must come off the diuretic immediately. All in a day's work for him, one might say: how was *he* to know the death knell he had just sounded to my only hope of escape from the near-hell in which I was living? Not long afterwards a doctor (it may have been my GP) told me that I must now be prepared to live with the condition for the rest of my life. But how *does* one prepare for what was for me such a catastrophic eventuality? Many would have said, perhaps even a little impatiently, 'Just *forget* music, and get on with the rest of your life.' *Their* problem, of course, was that music had never meant anything much to them in the first place, so forgetting it was - as seen by them - no problem at all.

For myself, however, I was to discover that I was entering a 'Dark Night of the Soul', as Saint John of the Cross, a 16th-century Spanish Carmelite monk, called it. At times harsh in the extreme, and tragic, it is not, however, a negative experience. It has been said of it that, 'Rather than resulting in devastation, the Dark Night is perceived by mystics and others to be a blessing in disguise.'

Chapter 28

The debacle that followed the ENT specialist's attempt to cure my cacophonous hearing by means of a diuretic felt like a very long snake indeed, taking me all the way back to Square One. Furthermore, there seemed to be no prospects remaining, that I would ever be able to move away from it again: yes - the Dark Night that I was facing looked a very dark one indeed. Even so, in my quiet times at the beginning and end of the day, there were moments when I was able to take comfort from the recollection that I had been there several times before in my life, and had learnt (like those mystics at the end of the last chapter) that provided I 'stayed with it' there could, and *would* be a light at the end of the tunnel – difficult though it was to believe at the time - a light which would reveal unexpected truths *and* blessings. I found myself asking, 'Could *anything* be darker than that moment when - clutching a poem envisaging a future with Mary spent enjoying the new freedoms that middle age would bring - I found her in a coma, and near to death?' Yet a new kind of life, undreamed-of in the poem, had come to us - a life which, during the twenty years that followed, acquired a whole new dimension and meaning which would have passed us by, had those twenty years been 'sunshine all the way'. Could history repeat itself, despite the perniciousness of the blight that had struck me?

I decided to embark on what has been, over the centuries, a time-honoured Quaker activity – that of keeping a 'Journal'. This is a much bigger, and more serious undertaking than what most people refer to as a 'diary', and concerns much more than the events themselves put on record in this way: it is none less than a day-by-day attempt to elucidate their spiritual significance.

It has been said (in Wikipedia) of the Journal of George Fox, the founder of Quakerism, that 'it is known even among non-Quakers for its vivid account of his personal journey'.

And of John Woolman's Journal (a contemporary of George Fox):

'it is considered to be an important spiritual document, but also a classic in English literature, as shown by its inclusion in the *Harvard Classics*. It is reportedly the longest-published book in the history of North America other than the Bible, having been continuously published since before the 1776 revolution.'

It goes without saying that the Journal *I* decided to start in September, 2003, though of the same *intrinsic* nature as such Journals, was not in the same league - and was, in fact, never intended for eyes other than my own. Perhaps its purpose was summed up in just one sentence at the very beginning: 'A balanced view of just what is going on is of the utmost importance, and will be helped by *reviewing* these notes from time to time.' I stopped writing the Journal after all but three years, for reasons that will become obvious - but which I mustn't anticipate - and during that time filled eight hard-bound A6 notebooks, considerably more than 100,000 words in all. It is well-known that the very act of attempting to write down what is going on in one's life can, in itself, be psychosomatically therapeutic, which is just as well, as a glance at the notebooks has convinced me that such is my scrawl that even *I* would find the bulk of them well nigh illegible now!

One thing, however, of which that glance has reminded me, is the writings of H A Williams and Monica Furlong (among others), which were an absolute lifeline to me at the time. There was, for example, Monica Furlong's account (in 'Contemplating Now') of the experience of monks subjected to the discipline of being required to remain in their cells for many weeks, virtually in solitary confinement. 'What,' you may well ask, 'has that got to do with hearing that has gone cacophonous?' The connection is by no means as tenuous as you might think. Whether you were able to enter into my feelings when I said that I felt I no longer belonged to the human race depends on how much serious music means to you, but I repeat: I am *not* exaggerating how I felt - and feeling that one no longer belongs to the human race has a great deal in common with how a monk feels after weeks of solitary confinement to his cell.

According tp Monica Furlong (and I shall be quoting her at some length), the monk feels threatened by the loss of his own identity, and, 'sliding down the precipice into nothingness' [as he sees it], he clutches at anything to prevent it. As Monica Furlong puts it,

> 'He must let go [of his old identity]. He longs to let go. He is terrified of letting go. He dare not let go. He cannot escape until he does let go.' She refers to the "*akme*", which, she says, according to Thomas Merton, is at once "the real point", and "the moment of truth". She continues: 'What is this "real point", or moment? It is the moment when the contemplative lets go. It is submission, self-giving, trusting (though none of these things are possible unless the conflict has been faithfully gone through). It is loss of [the old] identity. It is crucifixion. It is death.'

Strong words. Indeed they are, and they may be considered to be 'over the top', despite their figurative use. For myself, I did feel I was experiencing – in slow motion, so to say – that free fall into a threatened nothingness (remember, I had been told on good authority that I might have to live with the cacophony for the rest of my days). There were, as I have already said, tantalizing remissions for a few hours, and occasionally for a day or two, during which something like tunefulness would return, only to be marred by a strong 'boominess' or harshness, albeit confined to certain pitches, usually in the bass region. *Very* occasionally (every couple of months or so) the remission would be without blemish, and I would be in my seventh heaven, listening to as many of my favourite symphonies as I could, before returning to what I was coming to think of as my private hell.

How well I remember May 19th, 2003! (How is it that I remember the day and the date so well, when I hadn't even started my 'Journal'? Yes, indeed, thereby hangs a tale.)

The remission had lasted the whole day, and was with me, as (back on Cloud Nine) I clambered into bed. However, I had hardly snuggled down when the phone rang. It was Roger (or Wendy – I can't remember which) proudly announcing they had a second grandchild – Cloë – born that day to their older daughter Anna (who lives in Spain). I remember almost leaping out of bed, and going straight to my hi-fi in the sitting-room, and putting on a CD of a Jean Françaix wind ensemble – I can't remember which, but the majority of his music is boisterous and full of fun, and just right for a late-night celebration of the arrival of one's latest great-grandchild!

Such remissions - no sooner come than gone again - were very much a mixed blessing, providing, as I have already implied, a tantalizing taste of normality, but with no promise that they could, or would be anything other than fleeting. The joyous little episode just described took place eight months after that fateful night in the Autumn of 2002, but had little or no effect on the general gloom, and I succumbed to a deep depression, which in turn led to the worst bouts of insomnia I have ever experienced. My doctor rang the changes on nearly all the available drugs, usually with little or no effect. Well do I remember getting out of bed in the small hours one night, and, with a feeling that my legs just didn't belong to me any longer, I crashed into a corner of the room. Consulting my doctor the next day, I said I just didn't know what was happening to me. She responded, 'It's quite simple really. Your mind was awake, but your *body* was asleep.' So bizarre did this sound to me, that I resolved to try to wean myself off the medication rather than suffer such a weird and unnatural dichotomy between mind and body. And things did slowly improve, sleep-wise.

2003 wore on, and I was living from day to day. But, thanks be, the time didn't lag, for (apart from Monica Furlong) I began re-visiting the writings of those who, in the past, had similarly been of great help to me ('spoken to my condition', as George Fox would have put it) - most significant of all H A Williams, deceased now, but formerly a member of the Anglican Community of the Resurrection and Dean and Fellow of Trinity College, Cambridge. His best-known work, by far, is a collection of sermons called 'The True Wilderness', delivered to students whilst he was Dean of Trinity (Prince Charles being among them). In one of the sermons, called simply 'Gethsemane', the following passage occurs, which I shall quote at length, and which I came to know almost by heart during the summer and Autumn of 2003, so relevant did I feel it was to the crisis in my own life.

Having dealt most movingly with Christ's experience in the garden of Gethsemane, H A Williams asks,

> 'And what of us? It is unlikely that we shall ever have to go through an experience as deep and devastating as Christ's own in Gethsemane. But we may sometimes approach it, even if only from afar. Someone we love and on whom we depend may die. Or our material circumstances may unexpectedly change for the worse. Or something we had set our heart on and were working for devotedly may collapse into nothing. Or perhaps we already have discovered that our real enemies are inside us, that we have an unfortunate temperament in this way or that, that we are assertive or quarrelsome or timid or prone to worry and be anxious, vaguely but disturbingly frightened of something – we don't know quite what. Or perhaps we shall be ploughed up, turned inside out by a turbulent, unsatisfied love. If any of these things are true of us, or become true – and examples could be multiplied *ad infinitum* – then God won't provide a magic escape. If we look to Him to do that, we shall feel he has let us down. What God will enable us to do is to face these unpleasant, ugly, disturbing, frightening, sometimes agonizing facts or feelings. To face them without dodging or pretending that things are different, and then to accept them. 'This is me.' 'This is the corner I am in.' 'This is how I feel.' And, as with Jesus, so with us, the acceptance will bring the victory. By not evading our circumstances, outside and inside, we shall cease to be their victim and make them bring us to the very life which they would rob us of.'

H A Williams then goes on to tell the story of a man he knew personally, who was blinded in the Second World War.

'After about a year, he discovered that he could write music, although somebody else had to write it down for him. His music has not yet been published. Perhaps it never will be. But that is irrelevant. He wasn't looking for fame or reputation. He was looking for life. And he found it in the music he composes. A few years ago he said to me, 'You'll think it very odd, but before I was blinded my life was very shallow. I sometimes wonder whether I was alive at all. Now I have found a richness and a peace which before were unimaginable. Of course being blind is still hell, but I have learnt to live with it, and the privations it brings. And had I not been blinded, I don't think I should ever have discovered the deep happiness I now possess underneath the pain.' H. A. Williams concludes, 'This man had been with Jesus in Gethsemane. He had mastered his fate by accepting it.'

The soldier's eyes could no longer see light. My ears could no longer hear music, as music. He said it was hell, and I wouldn't disagree. After some time he had found a new basis for his life – music. Music had been the basis of my creative life, and I had lost it. Could I, too, hope to light upon a new basis and outlet for my creativity? Yes, was I, in fact (to return now to Monica Furlong) approaching my own particular *'akme*? –

'[the] death or moment of truth [that] achieves a transformation which is at the heart of the contemplative experience. In Merton's words, "We discover that *where we are is where we belong."* [Italics mine.]

'In mysterious fashion,' Monica Furlong continues, 'the pain and sense of loss, when they have been fully experienced, flower into new life. Pain becomes joy. Frustration becomes fulfilment. Death becomes resurrection. A sense of order and harmony floods through all our emotions and all our actions. Fear, boredom, lassitude, sadness, loneliness vanish.'

No – I was not approaching anything at all as comprehensive as that, but there were signs of a significant change – the *beginnings* of a transformation. Monica Furlong then goes on to a very important addendum:

'It is important to remember that though this experience is one more familiar to the contemplative than to most men, it is by no means an uncommon one among people who do not regard themselves as contemplatives. Many people have had this experience in recovering from bereavement, illness, marital and other forms of conflict, or, indeed, most forms of severe pain. *They are led to the puzzling discovery first that suffering and joy can be difficult to distinguish, and then to the suspicion that they may turn out to be the same thing.*' [Italics again mine.]

Gobbledygook? You can be forgiven for thinking so. 'After all', I hear you saying, 'who in their right mind would associate joy with *suffering?*' We need to be clear about what we mean by 'joy': for myself I make a distinction between 'joy' and 'happiness', even though all the dictionaries I have consulted make little or no distinction between the two. Typically, Collins English Dictionary states that '*Joy* is a deep feeling or condition of happiness or contentment,' and '*Happiness* is [the state of] feeling, showing, or expressing joy.' In contrast, a tiny poem I wrote in the wake of Mary's first stroke has it that

> Joy is no joy, which
> needs to state its terms,
> and so deny its own true nature –
> which of such substance is,
> that neither time,
> nor chance,
> nor circumstance
> erodes,
> or ever can.

For me, happiness is to be thought of as more ephemeral – more dependent on circumstance: joy runs much deeper. After surviving 95 years, I don't have the difficulty that some might have with Monica Furlong's statement that '*suffering and joy can be difficult to distinguish, and may turn out to be the same thing.*' You see, it is my experience that it is the apparent *meaninglessness* of suffering that makes the pain (whether physical or mental) seemingly unbearable. When it acquires significance and meaning in the eternal order of things, there is 'joy in Heaven' one might say, and joy in one's own heart, and yes, indeed, the joy and the suffering *can* then be difficult to distinguish: *meaning* is the Holy Grail we all seek in our lives.

So – if music was what changed the soldier's life, what was to take its place in mine, and similarly change my life? One morning, after breakfast, in October 2003, whilst washing-up, I happened to look up and glance out of the window overlooking the road. On the opposite pavement was a young woman, who was - somewhat wearily, it seemed - pushing a pram up the incline. It was as though I had taken a photograph of her, capturing every detail of what had so suddenly presented itself to me. And it was not just the visual details – it was the psychological aspects of the scene that had insinuated themselves. I dropped my eyes, thinking to get on with the chores, but found my mind and thoughts were no longer on the job: they had remained with the young woman. And almost before I knew what was happening, I had wiped my hands, grabbed a pen and paper, and was sitting at the kitchen table – yes (did you guess?) - *writing a poem.*

So what was so unusual about that? After all, as we have already seen throughout these Memoirs, it *did* happen from time to time – on average about once a year throughout the last 50 years of my life. The makings of an answer to my own question are in fact embedded in the poem itself:

>Leg-weary,
>lack-lustre, as she
>pushed the pram –
>was it homeward,
>to seek solace in a
>long-awaited cup of tea? –
>or was she outward-bound
>for the Supermarket, and
>the daily need to balance
>the price tags against
>a whole family
>to feed?

> Work-weary,
> she made the pram
> (the baby, too) look like
> an encumbrance she could
> well have done without –
> the transition from
> live baby to dead weight
> insidious; the joy,
> the miracle of giving birth,
> for time being quite forgot.
>
> World-weary,
> I watched with pristine
> vision late acquired from
> new things becoming old,
> and old things
> new:
>
>> the baby seen as though
>> first-born of all – first *ever*
>> to grace this Earth of ours;
>> a myriad of lifeless
>> protons, neutrons, and
>> electrons (and all the rest),
>> beloved of the physicist
>> for their obedience
>> to formulated rules, but –
>> in mother's womb –
>> preposterously
>> given life. Yes, *life!* –
>> to become the child
>> of which – rightly -
>> she would be
>> so proud:
>
> insidious, indeed,
> the transition from
> miracle, to mere
> mundane
> mood.

'World-weary, I watched *with pristine vision late acquired from new things becoming old, and old things new...*' My 'akme'? – perhaps, but not the sudden, all-embracing experience described by Monica Furlong. Rather was it an *inkling* of a process that was only just beginning.

Well do remember how it felt to be writing the poem, and the effect it had on me. Having, by then, for a whole year suffered music as nothing more than sheer cacophony (apart from those tantalizing remissions), there is no doubt that I had become somewhat more acceptive of what I most dreaded - the possible permanency of the condition. Of equal significance and importance, I was beginning to ask questions about the music itself that I had been writing. It was a joy to write, and great fun – a significant proportion of it, however, amounting to little more than the fruits of an unusual and highly pleasurable hobby. But, as a means of communicating to others the more serious things I had to say, it was restricted by their need to be able to understand and appreciate classical-style music: I was no purveyor of 'pop', 'punk', 'rock and roll', or whatever! There was one thing, however, that *did* give me some satisfaction - that family, friends, and others seemed to get a measure of inspiration from my doing any such thing - rather than lolling in an armchair - as I approached 'my nineties'...)

So, how *did* I feel as I was writing the poem? Strange, is the short answer. At the same time, however, I felt it to be liberating. That is *how* I felt, but at the time I didn't know *why*, though it wasn't long before that began to be apparent. The fact was, that during the next few weeks I found myself writing a poem virtually every week. I say 'I found myself' doing so, because there had been no conscious resolve or intent on my part so to do: it just happened - their subject matter ranging far and wide (from Gardeners' Question Time to the dubious benefit bestowed on a beggar by offering him a cigarette): only one or two of them related to my own life and its overarching problems.

So - why 'liberating'? The answer was simple and profound: even though my cacophonous ears were still deeply troubling (just as the soldier said that his blindness was still 'hell'), nevertheless I was beginning to realize that I was no longer clutching at whatever came to hand that promised to save me from the loss of my old (musical) identity. And why was that? Because it was beginning to dawn on me that *a new and more significant identity, in the world of poetry, was beckoning,* even though – hitherto - poetry had been somewhat side-lined in my life.

It went further than that – and here we enter one of the deepest aspects of the spiritual life - that of *guidance*, and how it comes to us. 'What's the problem?', did you say? 'You were prompted to write that poem about the mother and her child, after which it was just a matter of "one thing leading to another". Where's the mystery in that?'

You're right, and that's how *I* saw it – that is, until I found myself asking a very pertinent question - a sort of thought experiment, I suppose you might call it. 'Go back to, say, 2001, the peak of your love affair with composition, when you were writing a new piece every few days, portraying almost every aspect of this human life of ours, from 'Dreaming' and 'Waking', to 'Up and Away', 'Out and About', and 'Home Again' (as one Piano Suite had it). Then ask yourself, *"What do you think it would have taken to force you to abandon all that?"'*

In 2001 it would have seemed a purely hypothetical question, and one which I would have regarded almost with contempt, and certainly not worthy of serious consideration. Nevertheless, pressed very hard for an answer, I believe I might well have said (and in the most scathing tone I could muster), 'I suppose, *if music ceased to sound like music to me,*' hoping to end what I regarded as a nonsensical discussion, with what I regarded as an even more nonsensical answer. I might add that, in 2001, offered poetry as an alternative to composition, I would have turned it down out of hand. Yet, in the Autumn of 2008 – seven years on – I published an Anthology consisting of nearly 240 poems, having, between 2003 and 2008, continued to write, on average, almost a poem a week, instead of, hitherto, one a year.

It might be said that I had had no choice: I had, in fact – without realizing it - *been* moved on.

Was this, then, a classic case of God moving 'in a mysterious way, his wonders to perform' ? That's a serious question, and one that I am still pondering.

Chapter 29

For those last few weeks of 2003, and throughout 2004, the poetry that I was now writing regularly felt more like something that had been *added* to my life, *overlaying* it, rather than basically changing it; that insight was to come later – with hindsight. And whilst it was true that I felt that *composing* music was becoming of less and less consequence to me, the ability to listen to it was just as important as it had ever been: it was a vital source of spiritual sustenance, and its loss would never be something I would just come to accept 'philosophically': I would need to be able to say, with Thomas Merton, 'where we are is where we belong,' and with the degree of acceptance that that implies.

It is evident from the Journal, that 2004 was a 'roller-coaster' of a year, if I may put it so mundanely. From time to time there were entries like the following (for a particular day in December, 2003): '[Suddenly] discovered my ears were very musical, and listened to Beethoven's 5th and 6th Symphonies, as well as "Beyond the Darkness" ' [my own song cycle]. Such entries, however, concerning those precious, fleeting remissions, were beginning to be interwoven with others, referring to the poetry. Thus, on Saturday, January 17th, 2004, I wrote, 'I do have to believe that poetry promises to be a more effective outlet for my creativity, and a more effective way than music of saying what I want to say to others - when all is said and done, my music was not saying anything very new: the poetry most definitely is.'

Nevertheless, this insight is followed immediately by a reference to a passage in H A Williams' 'True Resurrection', from the chapter on 'Resurrection and Suffering', in which he speaks of the two kinds of pain, namely 'the pain which belongs to the destruction of the limited self manacled to the past, and the pain which belongs to the birth and creation of a fuller self'. He adds, 'For our birth and creation involve pain – the woman in travail its inevitable symbol,' and goes on, 'But in fact when we receive our suffering and are willing to feel its pain, it is both sorts of pain that are hitting us in the face – the pain of dying and the pain of being born. It hurts when the manacles which chain us to the past are broken. And it hurts when by our experience we are opened up almost forcibly to the future. It seems like one single hurt, leaving us all too agonizingly aware of its destructive power, and almost totally unaware of its creative power, *save for the faint glimmer of undefined hope.*' [Italics mine.] A little later he concludes, '[That] first faint glimmer of hope is the impact on us of the Eternal Word calling us into our future, calling us into [new] being...'

I have confined myself to that single passage from the chapter on 'Resurrection and Suffering', but so important was the whole chapter to me, that I even recorded it on tape, so that I could listen to it anywhere and at any time (I had done this with whole books, in the past!). Significantly, it is followed in the Journal by a reference to two poems, both of which were apparently receiving attention at the same time: I did, indeed, seem to be embracing my new future, but the embrace was far from a steady one. Thus, the very next day (Sunday, the 18th) I was making the point that it was not just music that the problems with my ears were affecting: their distortion of the spoken word [a hearing-aid would merely have exaggerated it] meant that 'I had lost the comfort and sustenance I normally derive from our Quaker Meetings, and from ordinary conversation. …. In fact, the *whole* of life has been so badly marred.' I continued, 'I have to try to open myself moment by moment, minute by minute, and hour by hour to a real sense of God's in-dwelling presence, and the purpose, and dignity, and stature that that gives to this otherwise 'blighted' life and living.'

Again, H A Williams, in his book, 'Tensions', has a wonderful passage in which, over several pages, he gives some account of his own (developing) thoughts and experience of God's indwelling presence, which ends with the startling conclusion that 'to be truly ourselves is to be *lived by God*' (italics mine). When I first came upon those words I felt I had stumbled on 'the pearl of great price', the many facets to their implications for ourselves and our lives deserving of daily concern and consideration. Indeed, six years on, that is exactly what they get in my quiet times at the beginning and end of the day.

Back then, to the Journal... Turning the pages for 2004, the 'roller-coaster' nature of my spiritual journey through that year is still all too evident. Thus, in quick succession, there are such entries as: (Wed 11th Feb) 'I continue to feel more "up together", as though I have "rounded a corner" in a very basic way. The poem continues to grow. It isn't because my ears have done anything spectacular, although in town, and on the bus, sounds were much more natural. I [just] think I have "grown"! '

But the very next day I was writing, 'I desperately need more social life – I'm far too much on my own, and it leads to all sorts of unhealthy introspection. But what can I do about it? It didn't matter when I was writing "Michael's Notes", and the short stories, and all that music, but, of course, poetry is not an on-going. day-to-day, hour-by-hour activity. .... I do not have anything like a social life in which to "lose" myself (in the best sense of the word).'

Yet, only four days further on (Sunday, the 15th), I was riding another crest. The theme of Meeting for Worship had been 'Love', and I had taken my courage into both hands and brought it to a conclusion with the shortest poem I have ever written – and am ever likely to write: *'Love is in itself a prayer.'* In the Journal I noted that 'It was much appreciated, and I felt so much more "up together" for having contributed significantly,' – (a noteworthy comment, with its implication that I had been feeling that I had less and less to contribute to the life that was all about me, and to which I still felt I barely belonged any more.)

And now a big change of topic: it is high time I mentioned 'Shiatsu'. It will be recalled that in the Spring of 2002 I had developed what was diagnosed as the relatively innocuous condition of 'glue ear' in my right ear. A friend had recommended that to hasten recovery I should take a course of Shiatsu therapy, saying that it was good at 'shifting lymph around the body'. I decided to 'give it a go', and arranged to attend Bath Shiatsu Centre. There I met Mercedes, one of the practitioners, and arranged to attend fortnightly. I have to report that my 'glue ear' proved to be too sticky shall we say, to show any significant improvement before succumbing (together with the left ear) to the catastrophic changes that took place on that calamitous September night. Was I disappointed? Strangely (you might think), I wasn't. You see, by then, and quite apart from the Shiatsu, Mercedes had established herself in my life as a very wise, and much-valued counsellor: hence my Journal entry for Tuesday, Feb 17th, 2004: 'Mercedes was wonderful. I felt she was praying as she worked, or that her work was a prayer in itself. I spoke of the difficulties of returning to normal, "ordinary" life again, after the last sixteen months. Mercedes spoke of the difficulties, after being "broken". She is wonderful as a counsellor, and as a therapist. She seems to have been where I have been, herself. I felt enormously benefited by the session - and for the rest of the day.'

Later, on the same day, poetry gets a mention again: 'Have started another poem, "The Christmas Tree".' [Then, immediately, some doubts.] 'I do wonder now whether I'm putting too much into, and relying too much *on,* poetry. …. I long for *ordinary* life again, and simple day-to-day contact with my fellow human beings. …. My left ear was ringing fiercely with tinnitus when I went to bed, and I feared I wouldn't get to sleep because of it. …. I did my best then to see the tinnitus as a tiny – *infinitesimal* - fraction of Christ's suffering, praying to be able to feel that He was with me, sharing the tinnitus - and I must have gone to sleep quite quickly then.'

The switchback continued. Only two days later, and I was noting that the tinnitus had subsided to 'a very faint whistle', and before the end of the day, apparently, I had listened to 'a really wonderful-sounding' Grieg's Piano Concerto, the 15th Symphony of Shostakovich, and, for good measure, Mahler's 9th.

On Friday, the 20th, I wrote, 'I believe that Mercedes gave me her whole self on Tuesday, and "set the stage" (i.e. provided the opportunity and the conditions) for Wednesday and Thursday.' And the wonderful remission for which she seems to have been largely responsible continued: the next day (Saturday) I listened to Stravinsky's 'Rite of Spring', Shostakovich's 5th Symphony. and Mahler's 5th. *What a feast!*

And, to cap it all, I spent that evening putting the finishing touches to a long poem called 'Grace'.

Maybe it should be included here as some indication of how things were going on the poetry front.

------------ **Grace** ------------

Words work through
meaning – shallow or
deep, edifying or
cheap – all but
self-evident in some,
others requiring the
services of a dictionary –
even an encyclopaedia –
to elucidate their full
import.

There are words, though,
which have no roots deep-
buried in the past, their very
sound synonymous with sense,
like "clip-clop", "hiss", "buzz",
"bang" and "plop".

Ironically – perversely, even –
they are categorized by a
word, six-syllabled, no less,
and one whose meaning, on the
face of it, could be hardly more
opaque: "onomatopœic",
for goodness sake!

And for Goodness' sake are
other words – most precious these,
of all – their etymological pedigree
impeccable; and, far from
onomatopœic, they yet
speak to us most plainly
through the beauty of
their sound.

Of such words is "grace" –
many-faceted, it's true, from
His Grace the Duke of Wherever,
to a prayer of thanks for food;
from extra time to pay a debt,
to a humbleness of mood,
and even "grace" of movement
which is effortless
and smooth.

But, it is as an attribute of
God, that the word bursts
bud, and blossoms
like the rose.

To say that God is
"gracious" is to take but
one small, stumbling step
towards an understanding of
what it means to speak –
often all too glibly – of
"the grace of God".
It is as though meaning
takes on such richness –
such fullness – that it
overflows into the very
sound itself.

Monosyllabic, and
mere five letters long,
it speaks of love without limit,
and love that sets no bounds –
a love that comes to meet us,
greet us, whilst we are yet
a great way off. There is
forgiveness, too, implicit
in this wondrous, five-lettered,
monosyllabic sound –
forgiveness that
states no conditions –
save one:
our readiness,
our willingness,
to accept.

Such love, and
such forgiveness,
are like the two halves
of a critical mass of
pure Goodness, which,
coming together
in the soul, generate,
with explosive joy,
new life,
new love,
new being.

The poem had been requested (my first ever!) by a Quaker friend, and I was able to deliver it to her at the Sunday Meeting for Worship, two days later. She was delighted with it! - saying that she would like to send it straightaway to a priest who had conducted a Retreat she had been on the day before. In the Journal I commented, 'I am finding a new identity (or part of it) in my new poetry. It's wonderful, and I am beginning to feel that this terrible episode of my hearing problems has been "used" to turn my creative energies from music to poetry. [But] I do hope still, to write *some* music.' This was more than an inkling of the 'akme' – it was a tiny 'taste' of the akme itself – the 'real point', the 'moment of truth'. But it was to take the rest of 2004 for me to recognize fully that I had indeed been 'moved on' – that writing poetry had displaced composition as a vehicle for the sort of things I wanted to say to my fellow humans.

It would not be appropriate to continue with detailed examples of the Journal entries throughout 2004 in order to illustrate this metamorphosis – my akme 'in slow motion': a few more crucial examples must suffice.

Sun 29th Feb: 'Before going to Meeting I had had an amusing experience in the bathroom, which produced a saucy little poem, 'Waterloo'! How much it shows the rise in my spirits! – and how much I enjoyed reading it over the phone to Hazel, Mary, and Pippa during the day!

'My ears are registering truer and truer the "timbre" of the piano, and I was delighted to find, quite suddenly, today, that I could pick out the F above middle C as *out of tune* – really 'twangy' to *both* ears! – and I hadn't noticed it before today. Amazing! (Fancy being *thrilled* to be able to detect that my piano was *out* of tune...) The poem "Litter" is already my longest, and is going to get *very* long.

'It's all very wonderful, this change that has come over my life these last 11 days.'

And yet, and *yet* - the entry for the very next day (Mon 1st March) began: 'Before getting up I discovered my hearing –

particularly the left ear – had deteriorated significantly. It was a bitter blow. I got very depressed trying to set this off by some hard work on "Litter".' Month after month passed, with further remissions from time to time, but none of them approached the eleven days of normality I had experienced at the beginning of March.

Significantly, however, there were Journal entries clearly indicating that I was beginning to address that very matter and my unhealthy preoccupation with it. Thus, in the entry for Monday, March 8th, I gave myself a right old dressing-down: 'Not being able to [listen to music], just when I would like to, is not so important as letting it ruin everything else I might do with my time and my life.' Then - in capitals! – 'GET ON WITH THINGS, LLOYD: SOME MORE *WRITING* ON SOMETHING OR OTHER (e.g. the "Memoirs", or whatever) *apart from* the poetry - which can't be written to order. Don't allow the situation to "dissipate" you.' [A slightly odd choice of word!].

On Friday, April 2nd, there is reference to a visit to Mercedes, for Shiatsu. '[Among other things] I also shared with her the beginning of the new poem [the first two lines of which were a slightly daring recasting of one of Wordsworth's most famous lines: "I wandered lonely in a cloud,"]. We had a very deep session, and even as I write, the very recollection of it fills me with sudden hope and optimism and a tinge of real joy. Wonderful!'

That reference to 'the new poem', with its mention of how it begins, is of considerable significance, and has just provided me with an idea for the future - of a 'Poetry Journal', recording the circumstances in which a poem came to be written. You see, although in the case of many of the poems which I published in the Anthology I could still remember the circumstances in which they were written, there were many others whose roots and origins had been lost – irretrievably, I thought. The poem discussed with Mercedes was called 'Wind from Russia', and – set in its proper context of April 2004, it assumes a much deeper significance at the *personal* level. Here is the poem:

Wind from Russia

I wandered, lonely, in a
crowd, my body close to
home, my spirit far away,
tossed, hither and thither,
on a sea of restless thoughts
whipped up by the winds of
change – threatening to
cast me, high and dry, on
a far-off, unfamiliar
shore.

    Suddenly, then,
    there came the sound of
    distant music, wafted on the
    morning breeze, bright, bold, and –
    yes! – *brazen.* (It was played by
    a brass ensemble.)

Jostled by the crowd,
I traced it to its source –
five men – *their* bodies also
far from home, but their
spirits (unlike mine just then)
had both local habitations
and a name:
two trumpets,
a horn,
a trombone
and a tuba.
What joy sprang
from the mere sight
of their gleaming
bells and valves, and
convoluted tubing!

    Their repertoire ranged
    far and wide. What matter
    that it was familiar, and
    sometimes near to trite?
    The wind from Russia had
    out-blown the winds of change,
    and my spirit – no longer
    storm-tossed and adrift –
    came home again, to music
    heard for what it was – voices,
    not just from Russia,
    but from Heaven
    itself.

Now it is obvious from the events depicted in the poem itself that my ears were plainly in remission. and yet, in the entry in the Journal, and in my discussion with Mercedes, there is no reference to this - my concern being to share the latest *poem* with her. (We had shared 'The Christmas Tree', and despite the doubts I was still having about the poetry at the time, she had declared it 'sad, but profound'.) *Of course* I would have been *listening* to music as avidly as ever, but, as a means of communication, and an outlet for those inexorable creative urges of mine, composition had, as we say, finally been 'put on the back burner'.

A week later, after the session with Mercedes, it was Good Friday, and, according to the Journal, I went to the Abbey (as I had done for many years) to participate in the Three-Hour Vigil, which, as I put it, 'I managed — but only just!' It set me pondering anew the whole problem of suffering — its causes, and our response to it. Inevitably, I was soon back again with the severe emotional trauma inflicted on me the previous Autumn by one I had thought to be a good friend, only to find that I was now castigating myself for being so easily 'taken in' — for being such a 'soft touch' - even feeling myself to be some sort of a failure, because of it. To quote from the Easter entry in the Journal: 'Far from feeling a failure because of it, I need to see it in the same light as (but only the tiniest fragment of) *Christ's* sufferings, which *He* also must have seen as the calamitous response to proffered good. His was the only answer to that kind of evil - to SUFFER IT CREATIVELY. May the truth of Easter (which those three words epitomize), and of this *particular* Easter, enable *me* to do just that.'

Within a matter of weeks, I had written one of my longest poems to date. It was called simply 'Active Service', and having dealt briefly with the traumatic incident itself, the poem continues:

A new, and fragile self
emerged – like a phoenix
from the ashes – to find that,
unwittingly, it was in the
front line of a battle,
ongoing since human life
began – its objective,
to wring out of suffering
a meaning commensurate
with the existence of the
rings of Saturn, or the
dimensions of the Milky Way.
It was a commission which,
more than willingly, I would have
relinquished – simply to
find myself back in Blighty
again.

All too soon,
I discovered that – within
the system of governance
that covered such bulky items as
Saturn and the Milky Way,
and, yes, suffering, too –
no arrangements had been made
for conscientious objectors,
less still, deserters. So,
like it or not, I was in it
'for the duration'. Worse still,
there was no prospect of
going on leave: one was,
perforce, required to engage
in combat 'round the clock';
no truce either, on Sundays,
Bank Holidays, or the days
set aside to celebrate the
lives of saints – in fact,
'no respite', the
order of the day.

This was the *real* problem that I was dealing with throughout 2004: the problem of *suffering* - of which my recalcitrant ears were just the form it took for me, at that particular time. It could have no easy, less still facile, answer. Yet, in the second half of the poem, I asked some of the most difficult questions raised by suffering, and the answers seemed to come:

The battle well and truly joined,
I found still one more question
demanding of attention –
time-honoured, this one:
'Why, for goodness' sake (yes! –
indeed, for *goodness'* sake),
does God allow suffering
at all?'

It's the question that the
atheist can neatly sidestep,
leaving it squarely on the
believer's plate.
'Where *was* God,
and what was He *doing*,
when the suicide bomber
blew himself up, with
an unknown number
of others?

'For that matter,
was he *really* auditioning
a new angelic choir (as I had
been tempted to surmise)
whilst that child was being
savagely beaten and battered
to death? And – much nearer home –
did he have his back turned,
so that he merely *heard* that shot,
[the shot my friend fired
into the heart of my feelings]
oblivious of its consequences,
for me?' The answers
needed to be good.

For many months
they were anything *but*:
'Suffering was the price
I had to pay to wipe the slate
clean of past misdemeanours' –
or – 'It was simply a process
of toughening up, like work-
hardening a piece of steel';.

Or, again – 'Life was still
evolving – there was bound
to be suffering "on the way" '.
Whichever the case,
it amounted to a God
absent without leave
from His creation –
and, for that matter,
absent from me,
in my moment
of direst need.

Strangely, then,
at those very moments –
moments when, of a certainty,
it would seem I should have been at
my most desolate, my most forsaken –
I came to feel, indeed, to *know,* that
I was not alone: that I was *not* adrift
in an empty sea, but a crewman
on a boat called 'Me',
with sure hands
on the tiller.

Trite?
Trivial?
Vintage Victorian
sentimentalism?
Then think again –
the implications providing
nothing less than the answer
to the question over-arching all:
'Where was God when the
suicide bomber blew himself up,
with who-knows-how-many others?';

so rooted and grounded
is each and every human life
in Him, that, yes, He was *there* –
being blown up
with the rest.

And the baby?
He was there, too –
beaten, battered, and
crucified,
anew.

> And that bullet
> through my heart?
> Yes – that's right –
> it went through
> His heart
> too.

Remember, this was only halfway through 2004, and my damaged ears were still very much with me, yet it would seem that the dawn was beginning to break on my Dark Night.

But I anticipate, and must return to Easter Monday. Bank Holidays (so called) can be quite a problem when one is living on one's own, especially if one's movements are restricted. The world and his wife seem to be out and about enjoying themselves, whilst you are stuck at home, doing your best to entertain yourself – or that is how it can seem, so often. Thus the Journal entry for Easter Monday was five pages long. I can't say it was 'much ado about nothing', but it was certainly much ado about spending so much time on my own. 'I do "hang in there", and *must* continue to do so. ..... I do have my moments, but how to turn moments into minutes, minutes into hours, hours into days? I long for life to take on the aspects of victory, rather than a long-drawn-out rearguard action, which is what it feels like so much of the time – that is, except when I am writing poetry. But, as I've said before, I can't do *that* to order. I have to comfort myself with the fact that God does enable me to *live* "normally" (more or less), even though it's not accompanied by "normal" feelings – I just have to do without those, at present.'

As I continue to turn the pages of the Journal, it is evident that there were quite a number of different threads to life at that time, which, willy-nilly, and somehow or other, I was managing to weave into the fabric of what would appear to casual acquaintances as 'living normally', but the threads so diverse (from writing poetry, to making the most of the hearing remissions whilst they lasted) that it was often difficult to keep up even an appearance of normality. The Dark Night was of course the spiritual backdrop to it all, but in addition there were practical problems and difficulties, not the least of which was the ever-haunting ogre of sleeplessness. So, at the end of April, the Journal tells me that I took myself off to my doctor, a visit to whom was always a tonic in itself: forget the prescriptions.

'She was very helpful and encouraging. saying that there were many people needing me, and made a special point about the poetry, saying she thought that the last 18 months was worth suffering, to effect the changeover to poetry.' She could hardly have put it more strongly, knowing full well, as she did, what I was going through. [From that very first poem of 'the new era' – 'Mother and Child' – she had requested that I emailed each new poem to her, 'hot off the press'.]   I left the surgery head high again, As a GP, she certainly knew all about suffering in its multitudinous forms, and she had held nothing back in sharing my own very special, and private agonies.

'But,' I hear you say, 'as a GP didn't she have to refrain from getting too involved?' In many – perhaps the majority – of cases that would be true, but, you see, she is a deeply spiritual woman, and thus has unusually deep resources to draw upon, in dealing with the wide-ranging demands made on her, in her work. *What* a blessing it was, during my own trials and tribulations, to have such a GP, and such a wise friend as Mercedes! – both of them going far beyond the normal professional limits in their response to my needs.

The 'see-saw', 'switch-back', 'roller-coaster' of a life – call it what you will – that I was having in the middle of 2004, continued, largely unchanged and unchanging, for most of the remaining weeks and months of the year. The poems continued to 'arrive' virtually on a weekly basis, though I must hasten to emphasize that this was 'on average': a particular poem might take a fortnight to finalize, whereas another might be completed in a matter of hours. Either way, by the end of the year poetry had definitely become my 'voice', a richer, deeper, and more serious one than my 'salon' pieces for piano could ever have been.

Strangely, and seemingly inexplicably, the Journal ceased abruptly on Dec 16th, and was not resumed until October, 2005. There were, as we shall see, tragic reasons for that. Nevertheless, so far as the hearing problems were concerned, I continued to make (quasi-clinical) notes until Saturday, Feb 5th, 2005, when they, too, came to an abrupt end, again for reasons which will become all too apparent.

The scientist in me had invented a range of hearing-level and distortion tests, using my cordless phone, and a whole set of symbols to cover everything from degree of deafness, to tinnitus levels and pitch, as well as 'drones' and resonances. A typical entry would look like this: 'Audibility abt. 5 cm (*sic*). Main voice, slight distortion. 1571 voice, slight distortion. Tinnitus – VVLLHPW, VVLL low-pitched murmur, each ear. No HB. ....'

Assessing these records over the ten days or so preceding Christmas, it would seem that in general there had been some significant improvement, which was *tending* to be maintained, but was still subject to fluctuations of various kinds. And in the few days immediately before Christmas, there were high hopes and expectations beginning to be expressed – but it would have been simply with a longer and even better remission than usual in mind.

Friday, December 24th – Christmas Eve – duly arrived, and, as best I can, I will reproduce the hearing assessments for the Festival period:

Christmas Eve 2004

8 am: L ear: main voice undistorted, and in vol. approaching R ear. 1571 voice also practically perfect. Audibility a modest 1 cm only. MW certainly audible but well down on R ear. NB Music at 4 am [*sic!*] very very good.

1.45 pm: Audibility 2.5 cm or more! Main voice, L_ear, distortionless, and almost as good as R ear. 1571 voice almost *perfect!* MW good, and almost the same for both ears. Right ear tinnitus VVLL. VLLHPW. HB heard.

7.15 pm: Audibility 3 to 4 cm! Both ears undistorted.

9. 00 pm: As for 7. 15 pm.

11.00 pm: As above!

Christmas Day

10.00 am: All OK – as above.

Midday: Ditto.

10.45pm: Ditto.

NB. R ear 'drone' noticeable throughout the day, but perhaps a bit better during second half of evening.

The Notes for Boxing Day, and the day after, were very similar, whilst those for the following three days recorded still further improvement, partially coinciding with a bout of diarrhoea! My comment: 'Perhaps the latter helped!' – a not entirely capricious remark, since diarrhoea notoriously reduces body fluids in the absence of compensatory fluid intake (shades of the diuretic prescribed by the ENT specialist...).

This wholesale improvement was maintained throughout January (2005), the records becoming briefer by the day, until, for example, for Jan 24th, the entry covering the whole day was just an 'OK'! It must be reiterated, however, that throughout this period I simply took it that I was presiding over the longest remission that I had experienced to date – no more, no less. That it might be signalling my return to the human race just didn't occur to me, but looking again at those records covering the Christmas period, it is clear that that return journey, quintessentially, took place on Christmas Eve – for me a solemn thought.

As already reported, the records ceased abruptly on Saturday, February 5th, but that was simply because the state of my ears – past, present, or future – no longer mattered to me, by comparison with the catastrophe waiting to burst on the whole family, on the morning of Wednesday, February the 9th.

Chapter 30

That morning, after breakfast, I was sitting enjoying my usual 'quiet time', before embarking on the business of the day. I was meditating, with a brief passage from Father Andrew's 'Meditations for Every Day' very much part of the general background to my thoughts: 'In every trouble and difficulty, let us try to put from us everything else, and say, "Now I come to Thee". So all things will shape themselves according to His will, and the occasion, whatever it may be, will render Him glory, and bring to our own soul ineffable and lasting peace.'

The doorbell rang – a very unusual thing to happen at that time of day – and when I went to open it, it was to find my cleaning lady, Eileen, whom I shared with Roger and Wendy, on the doorstep - a lovely lady, and a joint personal friend of ours. 'But *Monday* is your day,' I was thinking, as my eyes lighted upon her: and, indeed, she wasn't in her working clothes - she was in her 'Sunday best'.

Without further ado she said, 'May I come in?', and promptly did so, without waiting for an answer. Then, turning towards the sitting-room, she said, 'Come and sit down, I have something to tell you,' doing her best at the same time to convey that her visit was very much a formal one. It was *so* unlike her, I was thinking. But - almost hypnotized by her manner - I obeyed, whilst she solemnly took up a position facing me. I actually remember thinking, 'This is all a *charade!* - she's probably won the lottery!'

She was speaking again. 'Your son Roger,' she was saying, in that official voice she had found from somewhere, 'Well' – then a pause – 'he's *died*.' It was totally beyond my ability to believe what she had said, so I rejected it out of hand, and found myself responding with something like, 'Come on, Eileen! - what *is* it you've got to tell me?' 'It's *that*. Roger died suddenly, this morning, as he was getting out of bed.' 'Oh, God, no. She means it,' were the words that flashed through my mind, as the reality of what she had said began to strike home.

I have no recollection whatever of the next half-hour or so – of how and when she came to take her leave: the next thing I remember is the doorbell ringing again. She must have been still there to answer it, and - of a sudden then, I found my doctor was sitting beside me, holding my hand. Presumably Wendy had rung her, and she must have given up her morning coffee break to come and make sure that I would be alright. She is deeply Christian, as are all her colleagues in the practice (there is even a little box for requests for prayer on the reception desk), and, over the years, and particularly during the two years of the Dark Night from which I was only just emerging, we had shared much, spiritually: there was little need for actual words between us. She knew where I was - and that sufficed. The one thing I do remember was sharing Father Andrew's words with her - in that moment of sheer desolation they were a sheet anchor: 'In every trouble and difficulty, let us try to put from us everything else, and say, "Now I come to Thee." ' And she and I did just that, as she continued holding my hand in silence.

Wendy must have been utterly distraught (I learnt afterwards that she had witnessed Roger die), and I have no recollection of speaking with her at all, that day. Indeed, I have no recollection of Eileen's or Dr McHugh's departure, or of having had any lunch - though I'm sure I must have done. I do remember, however, exactly where I was standing (by the dining-table) mid-afternoon, when - with a savage reality all their own - these words (you may call it a poem if you wish) came to me 'all of a piece', with no desire on my part, then or since, to alter even one of them:

> A huge hunk of me died today.
> I didn't have any say
> in the matter –
> it happened,
> just like that.

> I've still got two arms,
> two legs,
> two ears,
> two eyes,
> a nose that still smells,
> and a tongue that still tastes,
> but I am less
> one son,
> *I am less*
> *one son –*
> I am less,
> I am less,
>
> I am *less*...

John came to my rescue that evening, motoring over from Whitstable to be with me for a few days, as I struggled to come to terms with what had happened with such tragic suddenness. Well do I remember thinking – in the computer terminology of our day and age – that we are not *programmed* to deal with the loss of a *child*. The loss of *any* near and dear one is desperately hard to bear, but it seems to be written on our genes that, nevertheless, we must be prepared to face the loss of – yes - a *parent,* or a *spouse,* or even a *sibling*. But the loss of one of one's own child? That seems to go against the natural order of things. 'That's just not supposed to happen,' you find yourself protesting, and before you know where you are, one part of you is questioning the reality of what another part of you knows full well *has* happened: you are split right down the middle.

The reason for this is both subtle and profound. I cannot doubt that Roger's *body* has died, and if I were having any difficulties over *that,* a visit to the mortuary would have been sufficient to convince me. The real difficulties arise from the fact that for the whole of his life - for those who knew and loved him so dearly – 'Roger' meant far more than a mere live body to us. In fact, of course, when you were with him, it was Roger the *person* that you would be engaged with, and to whom you would respond with joy and affection. And provided his body wasn't sick, you wouldn't give it - as a mere *body* – even a second thought. Thus it is that you find yourself facing the crucial – indeed *ultimate* – question, 'What does it *mean,* to say that *Roger* has died?' As I have said, there is no problem in believing that his *body* has died. But even an atheist – if he is honest with himself – would have to admit that in a deep and loving relationship much more is involved than two *bodies*: there is something intangible, something unique, something irreplaceable and of *lasting* value and significance to us, and which contributes to the very *meaning* of life itself.

You find that heavy going? Then what about the situation in which such issues, instead of being brought into the open, have been repressed, to rampage as 'loose cannon' in the subconscious? Mayhem can break out then in a bereaved family, and, sadly, it quite often does, anger taking the place of sorrow, and a desire to hurt replacing the natural instinct to console. And yes, you've guessed – it happened in our family. There is no doubt that Roger was the 'linchpin' of his family, and as such his sudden loss was catastrophic. He was a wholly loving and totally guileless man. All who were privileged to know him – intimately or casually – came under his spell. Little wonder that his unexpected, and so-sudden death created such problems for his family.

John and I, during the first two days of his stay, seemed to have an unspoken agreement between us to avoid - at least for the present - all the deeper aspects and consequences of Roger's death. In any case, there was a need to concentrate on the practical and more immediate matters, such as what items I would want to contribute to the eulogy. (It had been arranged that we would go to Wendy's on Saturday morning, where Roger's family was foregathering - his two daughters and their partners, together with the three grandchildren.) There was one thing in particular that I felt would be appropriate, and indeed of general interest to include, which was the intriguing way in which he came to choose architecture in the first place (recorded earlier).

On Saturday morning we duly made our way to Wendy's, where, sadly, but by no means unusually, the pain of sudden bereavement had, for the time being at any rate, proved too much to handle, and there was even anger in the air – some of it coming my way. In fact, despite my innocent enthusiasm for the inclusion in the eulogy of some reference to the intriguing manner in which Roger's choice of architecture as a career had come about, it wasn't long before I had to accept that there was to be no input from me at Roger's funeral.

A few days later, desperate to make *some* contribution, I got John to drive down to the funeral parlour, and on my behalf he placed a little 'stand-up' card on Roger's coffin for me. On it I had typed another little poem I had just written: I didn't sign it.

> "Thank you for coming," I said,
> the last time he called –
> I always did.
> And I always shall –
> "Thank you for coming,"
> "Thank you for coming,"
> "Thank you for coming."

After the funeral, I remember being asked in a rather crestfallen manner, by one of those present at that Saturday morning meeting, 'Did you write the poem that was on the coffin?' And I replied with a simple 'Yes'. By then, I had come to understand something of what had lain behind what happened that morning. They say that 'To understand all, is to forgive all,' but *that*, I had come to see, was only one aspect of the truth. And the other? – simply that in the circumstances we are discussing, 'To understand all is to realize that there is really nothing to forgive.' Grief had proved too great to handle, so it had turned into anger – something with which we are all much more familiar – more 'at home with': I know that's a rather crude way of putting it, but I'm sure you'll know what I mean. However, to begin with, it had been no ordinary anger, so that it was almost as unmanageable as the grief. It was anger against the very nature of life itself – that it should ever involve having to suffer such grief. It was anger against Creation. In the last analysis it was anger against God himself. And *that* was certainly too much to handle.

During those terrible days (and nights) before the funeral, I had my own mountain of disbelief to climb, my own dire need to convince myself that it was really happening – to fight off the perverse hope that it might yet prove to be some sort of nightmare (or should I say *day*-mare) from which I could expect to awaken, in due course. I wanted to remember Roger as he had been, so I had chosen *not* to pay that visit to the funeral parlour, which, literally and metaphorically, would have 'brought me down to Earth'. But that was not where I wanted to be just then: I did not want to be preoccupied with Roger's *dead body,* but with Roger as the *person* who had meant so much to me as my son for full 60 years, a son whom I could not think of as no longer existing. The implications of *that* for this human life of ours I would find quite impossible to contemplate: that the whole gamut of our richest experiences – all those things which, as we say, 'make life worth living' – no! make it *precious* to us – are an accidental and ultimately meaningless by-product of a totally blind evolutionary process. On the contrary, as I see it, the greater our sense of *apparent* loss (and I emphasize *apparent*) at such times as bereavement, the greater is the *meaning* and *significance* of what we have (*seemingly*) lost in the eternal order, of which our lives are a part. To compare the sublime to the near-ridiculous (in order to strengthen the point which I am making): the loss of a 'copper' coin would create *no* significant sense of loss in us, but to lose our wallet, with high-value bank notes in it, to say nothing of our cash card, would be another matter altogether. And to drive the point home in the harshest possible manner: how many high-value bank notes do you think Roger is worth to me? Just suppose I were to ask you that question in all seriousness. You would be shocked into speechlessness, wouldn't you? – dubbing me the most callous person you'd ever met, and deeming me to be – yes – *'sub-human'*. Interesting, isn't it? – considering we live in a world which, increasingly, tends to regard us humans, with the rest of Creation, as simply the work of a so-called 'blind watchmaker'...

Bereavement is not only an *event* – it is also the deepest and most challenging *experience* we have to face – that is, apart from our own deaths. The sense of loss that we experience is either *meaningful,* or *meaningless.* And, to sum up what I have tried to say: its actual *meaningfulness* is to be found in the very magnitude of the sense of (apparent) loss that we experience. I would be the first to admit that it is a very deep and difficult point I am trying to make, but it is vital to ponder it.

After writing those words, I realized that I had dealt with such matters (in more general terms) in one of the Lectures I gave to The Guildford Cathedral Religion and Science Group (see Appendix 3). In the lecture entitled 'A Scientist's Approach to the Resurrection', the following passage is to be found:

[In the crisis of 1947] 'it was given to me as in a blinding flash (which, as George Fox would have put it, 'spoke to my condition') that it was not *life* and death which were opposites and mutually exclusive, but *love* and death; that our deepest experience of love – of agapé, and all that is true and lovely and of good report in Eros – and of beauty, and truth, and goodness, and holiness, which are so intertwined with our experience of love - *all* these are at the very heart and meaning of existence itself; that the choice before us is a choice, on the one hand between death as the end of all, and on the other, of yielding ourselves to the evidence provided, not by our five senses, but by the quality of the response of the deepest level of our beings to these, the deepest experiences of our lives. Once we begin to respond at this level we begin to know that anything which superficially appears to make an end of love is itself a fraud and a usurper. And of course it is *death* which plays just this role in the minds of countless millions.   Love as we know it between husband and wife, parents and children, and in the richness of friendship, is just so much poignant nonsense if death is the *end* of it.

Now it is the sheer incongruity between death on the one hand and our deepest experience of love on the other that faces us all, ultimately, scientist and non-scientist alike, with the choice between living in the faith that the totality of our lives (which includes our life in the laboratory and our life in the home) is a *meaningful pattern, and whole* - and living in the last analysis *pointlessly,* believing that it is all 'sound and fury, signifying nothing'.   That is the stark choice......'

Such, then, was the complex spiritual journey which I took during those days before the funeral. I was pleased and gratified to find that the funeral itself had been organized as – basically - a *celebration* of Roger's life: his work, his art, and his sterling qualities as a family man and friend: the crematorium was full to overflowing.

There is something unique about the work of an architect: it is almost literally monumental, leaving behind him, as he does, structures in stone and brick, which may well live on for centuries after their designer has passed away. And thus it is with Roger. There are several projects of his dotted about Bath, and others in surrounding towns and villages. And almost the last was a Housing Association estate at Twerton, on the northwest edge of Bath. It occupies a gentle hillside, and is virtually one's first sighting of Bath as one approaches the City from the Bristol direction via Saltford. Typical of Roger, it incorporates various features aimed at avoiding the otherwise monotonous appearance of such estates. There are larger and smaller terraces – some of them gentle crescents, with a variety of roof-tile colours distributed among them. And in keeping with Georgian Bath, there are even modest chimney stacks, having, of course, a purely decorative function!

What a privilege and joy it is to be greeted by a bit of quintessential Roger as one approaches the City!

Here then, to bring this Celebration of Roger to an end, is the final one of the eleven poems I wrote 'In Memoriam':

Let there no misunderstanding be:
had he, in truth, received a knighthood
for his architecture,
or even been raised to the peerage
in recognition of his work,
and,
furthermore,
had become an Academician
for his painting –
had any,
or all,
of these
come to pass,
they would have counted for little
by comparison with his accomplishments
in the arts of living,
loving,
sharing, and caring for
all the little, precious things
that make up our human life.

Would all those who turned up
at his funeral have come
simply to celebrate the fact
that he was Lord Roger,
or even Academician Kemp?
No! –
peers are ten a penny,
academicians
three for the price of two.
But Roger?
There was just *one*
of him.

Chapter 31

Ten days after Roger died, and four days before his funeral, I started what might have become a new Journal. I called it simply 'A Journey', but I could equally well have called it 'Life after Feb 9th, 2005', for the whole family had moved into new, and uncharted territory. However, and again for reasons which I do not recall, it quickly came to an end - this time after barely a week. But, for me, it was a highly definitive week.

In the very first entry, on February 19th, the passage from Father Andrew's 'Meditations for Everyday' was still very much in my thoughts. I wrote, 'This morning, as I contemplated this passage yet again. a peace that I haven't known since Roger died came to me, as I found myself saying, "Now I *have* come to Thee." And in that new sense of peace there was also a sense of promise – that I would be given the strength to live again, in this real world of ours, to help and succour all my family and my friends, instead of it being the other way round.'

Directly after the funeral, John and Catherine took me back to Whitstable to stay with them for a few days, and on Sunday, the 27th, four days after the funeral, I made the following entry. 'After the most traumatic week of my life, I was standing in the bathroom after breakfast, when a wave of assurance broke over me: there *is* a future for me, with both meaning and purpose. It was one of those self-validating moments of truth.' And the next day, referring to the Sunday entry, I wrote, 'There has been no loss of those experiences of *eternal* value – in fact, in a sense, Roger's passing has *emphasized* and even *added* to them. We have, in fact, had to view life *from* the eternal standpoint, and many things are clearer, and more sharply focussed, than before. Again, his passing has made us more aware of what the human state actually is, in all its uncertainty. Actually, nothing has changed, apart, of course, from the terrible loss from our midst of Roger himself: we have been made aware of life as it has *always* been.'

The most likely reason for the Journal's short life was that poetry, in effect, took it over. Having written several hundred words a day for the better part of three years, I had, I suppose, become somewhat disenamoured of the sheer routine of it. Whether one felt like it or not, one felt duty-bound to keep it up, day upon day. I suppose, also, that the extraordinary way in which poetry had so quickly become part of my daily life was an indication that I had become less introspective, and more concerned with giving voice to the broader aspects of my thoughts and feelings, for which poetry was a much more appropriate vehicle than a Journal could ever be. Roger's death, and its repercussions for the whole family, would continue to lay claim on my time and energies in this way. Inevitably, over the months and years that have ensued, there have been further thoughts on bereavement itself: it can never be one of those matters which one can feel one has finally 'dealt with' in all its aspects, and which can therefore be tucked away in one of the drawers of one's mind. There was, for instance, 'Bereavement Revisited', written some two years after Roger's passing:

> In vain my eyes have sought
> the line of distant hills,
> from whose very sight
> I have drawn such comfort;
> the horizon, too, dividing
> what lies within my ken
> from that which lies beyond:
> the meeting place of Earth
> and Heaven vanished,
> overnight.
>   And in its place
> a drab-grey mist, stretching
> from Heaven's infinitude
> down to all that's left
> of Mother Earth – a lone tree
> by my house, doing its best
> to reassure me that what
> vanished overnight has,
> for time being,
> merely passed
> from sight.

I was moving on, and, more recently, wrote 'God's Alchemy':

> Consumed still, as I am,
> by such sense of loss
> and pain –
> yet, do I believe
> that time will be
> when pain
> becomes transformed
> into a kind of holy joy,
> and loss into
> mysterious gain.

There came the time when - whilst continuing to be concerned with such solemn matters as bereavement - an element of humour began to find its way into my poetry. With hindsight, I can see that this was a good sign, and possible evidence that I was beginning to come to terms with Roger's tragic death. Hence the Anthology, 'What, on Earth, Are We Doing?', was divided into three sections: 'In Lighter Vein', 'The Everyday', and 'The Spirit'. One quite unusual feature of the Anthology as a whole which I would like to mention is that in all three sections parallels have been drawn between the human experience, and the properties and behaviour of a computer! The comparison can at times be both an amusing, and indeed fruitful one...

So, 'In Lighter Vein', among other poems relating to the computer, there is 'Problem Solved':

> If it happens that life
> is not all that it has been –
> if you're finding it difficult
> to perform the simplest of
> tasks, and, even if you
> pull yourself together
> and manage to respond,
> it still takes aeons of time;
> if, worse still, you
> do the equivalent

of collapsing into an
armchair and closing
your eyes and trying to
forget the world in
sleep: if all this
should be the case –
then, take heart! All is
far from lost. Just peep
into your diary (so to say)
and find a day when
you made a note that
this would be one of those
that you would like to
return to, and, hey presto! –
in next to no time – all is
well: life can begin again;
and those seemingly
insurmountable
obstacles, that had
made it too difficult to
bear, simply not there
any more.

That is, if you're a
computer, with
"System Restore"
installed.

And, in the section, 'The Spirit',

Help from an Unexpected Quarter

My computer is helping me
to pray.
Surprised?
Yes – so was I!

> Of course, it has
> innumerable drop-down
> menus dedicated to
> solving particular
> problems (God has,
> too).

But there's something else
my computer claims to do:

it *"enables"*. Thus,
it tells me it has "enabled"
my Antivirus software,
so that it can carry out
its allotted tasks.

In the past,
I have asked
for God to
help me
(i.e. for one of those
drop-down menus
to appear); but now
(and much more
comprehensively)
I'm beginning to pray
simply to be
"enabled" –

and that covers
everything.

*　　*　　*　　*　　*

And *that* seems to be the appropriate moment to tell you that it is 6.13 pm, on Monday, September 28th, 2009, a nondescript, dull, early Autumn day with little, it would seem, to distinguish it from countless other days in my long life. But *that,* in fact, is not so. Roger, when he was a young boy, and the day had been rather special for him, would announce quite formally that 'Today has been a *main* day!' And today, if it lives up to its present promise, could be a main day for me. And the reason? – simply that I might well have finished these Memoirs – all 130,000 words of them – before I wend my weary way to bed.

So - is that it? Well - not quite. I thought that before drawing this journey through the past to a close you might like to know a few personal details:that at 95+ I am fortunate enough to be still in charge of my affairs – and by 'affairs' I don't just mean my bank account. I have lived on my own ever since Mary died in 1988, and 'do for myself' in almost all respects - apart from keeping the place habitable, for which purpose Debbie, my cleaning lady, pays a fortnightly visit to clear up the messes I create between-times... (Shall I let you into a secret? Actually, I myself have to do something of a clear-up *before she comes*, otherwise she wouldn't know where to start. In any case, pride would prevent me from allowing her to see the chaos in which I live from day to day: I have been known to describe the whole house as a 'paper chase' – with almost every chair a makeshift filing cabinet.)

To continue then: I still do practically all my own shopping, aided and abetted by a rucksack (it would be impossible otherwise), quite often returning home with ten pounds' weight on my back. (I've got the shop assistants well trained: I just turn round and lean back towards them, and they load it up for me (it saves hauling it up onto my back when it's packed full). Oh, yes – and I still manage to take a bath unassisted: I just go over onto all fours before standing up. Ah - I almost forgot: I gave up gardening some ten years ago. Jeff, a very pleasant young West countryman cuts the grass every three weeks and, as required, trims back the shrubs, thus preventing them from turning the garden into a wilderness rivalling the one I regularly create inside the house. And finally, perhaps I should add that before breakfast I do half an hour or so of the exercise regime known as 'Pilates' (after its founder, some fifty or more years ago). It's designed specially for 'oldies', to enable them stay in touch with their muscles and joints; a measured mile on my exercise bike then, by which time (and I hope you'll agree) I think I've earned my breakfast.

'Surely, that really *must* be it!'
No – not quite. There is just one more thing – and the most crucial of all: *what do I feel it has all added up to* – this

long life of mine, and still on-going? The short answer is in just three (very hackneyed) words: 'Love is all'. Their full meaning for me was embodied in a poem written in 1987 – over 20 years ago – called 'The Currency of Love' - probably the most significant poem I have written, and am ever likely to write.

>Love has a currency
>all its own:
>
>>its smallest denomination
>>is of inestimable worth –
>>yet it is thief-proof,
>>and needs no protection.
>
>Minted freely,
>It creates no fear of inflation –
>in fact, it reverses all the usual rules.
>
>>Thus, unless it is counterfeit,
>>its investment seeks no return;
>>and income actually increases
>>with expenditure.
>
>Its nature is always to be a gift:
>it cannot therefore be earned,
>or claimed as any kind of recompense
>or reward;
>and, when returned,
>it needs to be immediately re-invested,
>for the attempt to hoard it
>leads to bankruptcy.
>
>>It defies drawing up a balance sheet,
>>for it cannot be enumerated;
>>and any attempt to put a price on it
>>renders it worthless.
>
>It is always available on demand,
>and requires no security;
>indeed,
>it may take the form
>of a blank cheque,
>with the consequences
>willingly accepted –
>for counting the cost
>is foreign to it.

It is exchangeable
the world over,
but such exchange
must always be
person to person:
no broker
can have dealings in it
to achieve a cheap gain.

It can be taxed
to the limit,
yet emerge
with enhanced reserves.

It is the only currency
adequate to meet the cost
of living.

Afterword

What If...?

When I was young,
my elders used to say,
'Now leave it
as you found it' –
where "it" was a box
of my favourite toys,
a pile of neat-stacked
story books, or a drawer,
full of clothes, and destined
to serve as props, in a
game of "make-believe".

But, what if "it"
is a whole planet:
of grassy meadows
and tumbling streams,
of towering mountains
and rolling seas,
of leafy hedgerows
and blossoming trees;
of children who think it
will all stay the same
for them –

but it won't?

Despite having, seemingly, brought these Memoirs to a close at the end of Chapter 31, it came to me that my journey through the past was in need of an 'Afterword', as I have chosen to call it, which would not be about the past, but the future – the future about which I have grave concerns, yet in which I shall play no part. And those concerns, I realized, were epitomized in the poem above – written two years ago.

And the children? – they're children the world over, of course, and among them *my* children, and *their* children, and their *children's* children - my great-grandchildren, of whom I already have eight: Guido, Chloë, Paloma, Hannah, Daisy, Matilda, Béla and Delilah – as of this moment ranging from seven years old, to just a few weeks. So who is going to tell me that the world in thirty years' time is no concern of mine, even though I shall play no part in it?

There are many who still want to bury their heads in the sand, and pretend it isn't happening – that things are never as bad as the perpetrators of doom and gloom make them out to be – their own secret agenda, however, to defend their right to jump into their car just to pick up the morning paper before breakfast, or flit off by air to Paris – or wherever - for a long weekend; the Nimbys, too, who want to preserve for themselves the uninterrupted view from their bedroom window, rather than wake up to the engineered elegance of a wind farm whose gently-turning blades may help prevent the view from disappearing altogether, under a rising sea level caused by the melting of the polar ice-caps. Even more dangerous are the politicians, who have yet to bring themselves to take the decisive action so urgently needed, if life on Planet Earth is to have a future - allowing their decisions to be largely determined still, by their consequences for the next General Election, or the short-term economic well-being of their own particular country.

And what of those who put their faith in the wisdom of the Selfish Gene to ensure the future of the species? Just after Christmas, 2007, I wrote a poem about that too, called 'Thoughts post-Christmas':

"Unto us a child is born" –
parenthood, it's called,
but, terms and conditions
apply; although, these days,
it must be said, the small print
too often goes unread.

Be that as it may,
yet doubt not that
the small print is
getting bigger and
bolder by the day –
no longer possible
to ignore, or for

> parenthood to be
> restricted to fulfilling
> the dictates of
> the Selfish Gene.
>
> All too well now,
> do we know
> that safeguarding
> *our* offspring is
> no longer to safeguard
> the future of *their*
> offspring, and *their*
> offspring's offspring,
> from the consequences
> of a lifestyle that
> threatens survival,
> generations on.
>
> Make no mistake: we,
> the present bearers of
> an All-Too-Selfish Gene,
> needs must undergo a
> sea-change in the way we
> live, if the children of
> our children's children
> are, in fact, to
> live at all.

Yes – a 'sea-change' – nowadays signifying simply an *immense* change, but, as Shakespeare's Ariel has it, a change 'into something rich and strange'. In the poem, the change envisaged in our life style is certainly 'immense' – hence our reluctance to embrace it – but it is indeed also something 'rich and strange': 'strange' to 21st-century man because it is not self-serving, and 'rich' because it embodies a total, but deeply enriching spiritual make-over of our lives, and the way we live them.

The Quakers have a little book of 'Advices and Queries', epitomising their beliefs, and the wisdom which they feel has come to them over the centuries. There are just 41 of them, and the very last is as follows:

'We do not own the world, and its riches are not ours to dispose of at will. Show a loving consideration for all creatures, and seek to maintain the beauty and variety of the world. Work to ensure that our increasing power over nature is used responsibly, with reverence for life. Rejoice in the splendour of God's continuing creation.'

If, after my 95 years on this beautiful planet of ours, someone were to ask me, 'Do you have a creed that you would like to pass on, I might well refer them to No 41 of the 'Advices and Queries' which grace the early pages of our Quaker compendium, called 'Quaker Faith and Practice'.

BIBLIOGRAPHY

1. 'Contemplating Now', by Monica Furlong,
   Hodder and Stoughton, 1971

2. 'The True Wilderness', by H A Williams.
   Constable, 1965

3. 'True Resurrection', by H A Williams,
   Mitchell Beazley, 1972

4. 'Tensions', by H A Williams,
   Mitchell Beazley, 1976

5. 'Your Sort of Courage', by Lloyd Kemp, 1994
   (which, itself, contains a comprehensive Bibliography).

6. 'What, On Earth, Are We Doing?', and Other Poems,
   by Lloyd Kemp
   Aspect Design, Malvern, 2008

APPENDIX 1

CHRISTIAN STEWARDSHIP AND COMMUNITY

An experiment in community living by the Stevenage Peace Group

Our thought and feeling on the subject of community began to crystallise in the tense atmosphere of May, 1940, when the war flamed up in the form which we had contemplated before September 1939. Like everyone else who had been thinking deeply about it, we felt that it would bring with it some very big changes in our society.

We were at first inclined to believe that the changes demanded in our own lives and way of living would have to be quite fundamental: the temptation was to do something spectacular.

So it was that the first type of community which we discussed was the "withdrawn" type, involving the simplest of lives, in a more or less self-supporting group.

Of such communities, Middleton Murry has written:-

'The history of the manifold attempts at such self-supporting communities is not encouraging. Nevertheless, the effort is necessary; but it is to be conceived rather as a concrete criticism of modern society, and as an attempt to establish places of restoration and refuge from its anti-human stresses, rather than as an alternative form of society which could eventually replace the present form'.

Our Group agrees with this conclusion of Middleton Murry's. We feel that the good things which science and industry have placed at our disposal are not wrong in themselves, and that the problem of their manufacture under un-Christian conditions, their mal-distribution, or misuse, cannot be solved by abandoning them completely. It seems to us, too, that it would be an illusion for such a group to believe that it had dispensed entirely with the products and amenities of civilisation as we know it, since, for instance, such a group would presumably call in a doctor if one of its members were ill, and doctors are possible only within the complex framework of our civilisation. We do not feel that all men and women of this or any other generation are called on to give up the amenities of civilisation; and though with Middleton Murry we recognize the value of withdrawn communities as places of temporary refuge, we see in them neither the solution nor even a step towards the solution, of present-day problems.

Therefore we sought another kind of community, and next considered the type in which all the members live under the same roof, and though engaging in different occupations, pool their resources, the pool then being divided out equally.

What is it that is almost repulsive about so mathematically uniform and precise a scheme? Is it not that, for a start, it ignores the fact that different people have different needs and interests requiring different resources, as well as different tastes, needing different environments? Further, if such a scheme *were* possible among a group of individuals, it is still difficult to see how it could be applied to a group of *families* without affecting adversely life within the family. Those of us who thought specially about this aspect of community life, felt that family life is at the centre of the Christian concept of society, and that the family is a unit out of which a Christian society can and will be built one day. To dilute family life into group life would hinder rather than help the emancipation of civilisation. Once recognize the need of individuality in life, and above all in family life, and absolute equality becomes not only undesirable, but in some respects definitely impossible.

Convinced that we were not required to give up entirely the products of civilisation; equally convinced that we were not required to live under the same roof and have absolute equality in all things, we turned to consider Income Community experiments.

At least one attempt has been made to base such a scheme on the average income for Great Britain. Basically, it is argued, we should live at this level of income, and surrender any surplus to a group pool, to be used to make up to the national average those incomes within the group which fall below it. Any surplus remaining could then be administered by the group in the service of the common interests which brought its members together. However, even in the scheme to which we refer, the national average is by no means adhered to in practice. A statement of income and expenditure (based on necessities only) is drawn up by the individual, and submitted to the group. By "necessities" it is clear that the members include books, tools, etc., for their own work, travelling expenses, and similar extras. To be supplemented from the group pool, a member's income does not have to be less than the national average, but simply less than his expenditure estimated on the basis already described; and this expenditure may, and generally does, exceed the national average.

Now we have already made it clear that we in our Group feel that the aim of community is not necessarily the attainment of equality, and our urge to express our spirit of community was not to be satisfied by a scheme whose main principle was income *levelling*. Even in the ideal society, the diversity in natural abilities and needs of the individual members, to which we have already referred, must call for corresponding differences in personal expenditure; and whilst the national average should undoubtedly be borne in mind in making decisions regarding how much one should spend, or what one should regard as necessities, we feel that the ultimate justification for all expenditure must be that a sense of *stewardship* accompanies it. Particularly should this be so in the case of expenditure incurred which brings the amount above the national average.

Thus, for the first time we saw clearly the basis of the scheme for which we had been searching. For none of us had up to that time fully realized our responsibility for what we spent or saved. Was it more than enough to keep a healthy mind in a healthy body? If so, the surplus must be pooled and group-administered.

Now a healthy mind involves spiritual, intellectual, and aesthetic activity and expression, and we did not rule out as inessential expenditure on good books, for instance, or gramophone records of good music; and a healthy body involves wholesome food. Real inessentials can be avoided without going to the extremes of spiritual, mental, or bodily impoverishment. None of us has become convinced that because many thousands *are* impoverished in one or more of these ways it will benefit society if we add ourselves to their number. Also to be avoided in a true community scheme, is false economy, such as the purchase of shoddily-made articles, which at best represent bad craftsmanship, and at worst the products of sweated labour.

The individual knows that the Group expects that the decisions which he or she makes regarding 'necessities' will be honest and searching ones. The budget is then drawn up, the surplus - if any - calculated, and submitted confidentially to the one whom we may call the treasurer of the Group, who alone knows the individual contributions. We do not pass round detailed budgets. No one's conscience can be the judge of another's.

If the fundamental basis of 'necessities' is adhered to, and the budgeting done conscientiously, there is no doubt that in the average group, a substantial surplus will be revealed. In our own Group of nine, consisting broadly speaking of 3 office workers, 3 manual workers, and 3 professional folk, we have, over a period of ten months, had an average weekly surplus of nearly £3. We venture to suggest that this is considerably more than the average group of that number would at first think they could devote to non-personal expenditure. The scheme is, in fact, realistic, and does constitute the fundamental adjustment which we had sought to make. In the discipline both of the stewardship which it calls out, and the simple living which it entails, it is a vital preparation for life under the conditions which must obtain after the war.

Since the adoption of the scheme, the normal activities of the Group have been carried on, to which has been added social service work which the scheme has made possible.

Each member is free to place before the Group recommendations as to the way in which the surplus should be spent, and the final decision is made as a Group. The following are some examples of the way in which the surplus has been employed.

(1) Four poor people have been provided with half a crown's worth of groceries every week since the start of the scheme. This has been done anonymously through a local grocer.
(2) First aid equipment has been provided for each Group member.
(3) Quantities of coal have been purchased, which we have managed to get delivered anonymously to folk in need.
(4) Large quantities of wool have been purchased and knitted up into children's garments, some of it by Group members (including one of the men!), some of it by non-members willing to help. The finished garments, together with some purchased ready-made, have been distributed to poor children – especially the poorer evacuees – through the agency of the Sunday schools and day schools.
(5) The necessary materials have been purchased for the manufacture of about 20 mattresses for use in evacuation or air raid emergencies, and the Group is engaged in making them up.
(6) Substantial Group contributions have been made to certain distress funds, and to an East End mission.
(7) Group members have been helped during periods of unemployment. Books on religious, political, and social

topics are purchased as Group books from Group funds, and we are accumulating a little library. Also, general group activities, such as travelling to meetings outside our town, are paid for out of Group funds, a fact which avoids having to consider whether or not a particular member could afford it if we went.

Of all these Group activities we must emphasise that they are 'not an attempt to improve the looks of our own spiritual physiognomies' (as a Friend has put it) 'but a desire to draw out the best in others. We wish to create the conditions in which all men will naturally come to understand something of the search for the Kingdom of God. By simple living we may help others to break the tyranny of an over-rigid materialism'.
It will be clear that the scheme has brought new life to us, and new spheres of activity of great social significance.
It will have become evident to some that the principles which we have accepted are, broadly speaking, those on which the Third Order of St Francis was based. The close approximation of our scheme to the Third Order is best illustrated by this extract from John Hoyland's book, 'The Way of St Francis and Today'. Of the Third Order he writes:-
'Here we have something designed exactly for our condition …. the spiritual basis not merely of effective internationalism, but of the new, more Christian social order for which we all long. The principles are simple enough. A rigorously limited personal expenditure: the surplus of income devoted to the wisest possible service of humanity; manual work with our own hands on behalf of others, at any rate in all the time we can spare; the service of the needy, and especially the sick, again with our own hands …. definite work for the peace and unity of mankind; and the bond of an association, a fellowship, founded on religion – on a passionate adoration of Jesus Christ, on a resolve to take His principles and example as literally as is possible in view of family obligations ….

'Now on that most superficial observation of these principles, it will be clear that the Third Order was not something just for the early thirteenth century. It was a permanent thing, adapted to all ages, all environments, all peoples. *We* have got to reckon with it *now*. Very possibly our own civilisation is perishing of suicidal mania before our eyes just for the lack of the soul within it – this soul which inspired its supreme greatness of achievement in that wonderful century which Francis made.

'If you are like me, you feel the miserable inadequacy of any merely external, distant, sentimental adoring of St Francis; and therewith you feel the compelling urge to get down and *do* something about it. Cannot we have today a new Tertiary movement, binding together, however loosely, men and women of all manner of confessions and outlook, in a fellowship pledged, so far as in them lies, to these principles of 1220?'

Bibliography

1. 'The Defence of Democracy' by J Middleton Murry (Jonathan Cape).
2. 'Income and Community – Notes for Discussion' (obtainable from Patrick Wilson, 33, Townsend Drive, St Albans, Herts).
3. 'The Way of St Francis and Today' by J S Hoyland (SCM Press).
4. 'Simpler Standards of Living' (reprinted from 'The Friend', and obtainable from Friends' Book Centre).
5. 'Christian Stewardship and Community', an article by our own Group in Peace News, August 2nd, 1940.
6. 'The Challenge of Simplicity', an article in 'The Friend', March 21st, 1941, and subsequent articles in the series. 'What Friends Might Do Now'.
   Further sources are indicated in the books and articles to which we refer.

APPENDIX 2*

*Reproduced by permission of the Bridge Pastoral Foundation

Frank Lake, M.B., M.R.C.Psych., D.P.M., was one of the pioneers of Pastoral Counselling in the U.K. He was the founder, in 1962, of the Clinical Theology Association, and remained its guiding spirit up until his death. Its primary aim was to make clergy more effective in bringing the healing power of the Christian gospel to bear on troubled minds by helping them to understand and accept the psychological origins of their parishioners' personal difficulties. However, the training seminars in pastoral counselling, which he began in 1958, eventually enlisted professional and lay people in various fields, of both sexes, and of every denomination. 'Many thousands of people have passed through the seminars.

Dr Lake was born on 6 June 1914 in Aughton, in Lancashire. His staunchly Christian parents had deep roots in the Parish, where his father, John, a stockbroker in Liverpool, was organist and choirmaster. His mother, Mary, had trained as a teacher. Frank was the eldest of three sons.

He studied medicine at Edinburgh University, graduating with diplomas in medicine and surgery in 1937. Having long been attracted to missionary work, he then trained in parasitology at the Liverpool School of Tropical Medicine before taking up an appointment with C.M.S. to serve in India. As World War 2 went on, he was recruited into the Indian Medical Service, from which he emerged with the rank of lieutenant-colonel in 1945. His fiancée, Sylvia Smith, joined him in 1944, and they were married in Poona, where the eldest of their three children, David, was born. In 1946 Frank was posted to the parasitology department of the Vellore Medical Centre.

Dr Lake's momentous change of direction – from parasitology to psychiatry - came about after he was appointed Superintendent of the Christian Medical College in Madras. In the course of setting up a psychiatric unit there, he became concerned with what he called 'a variety of imponderable emotional factors which I had never been taught to think about seriously before'. Following his next furlough in England, in the early fifties, he undertook re-training as a psychiatrist, first at The Lawn, Lincoln, then at Scalebor Park Hospital in Burley, Yorkshire. His allegiance was to the 'Object-Relations' school of thinking, which considered the development of good human relationships the lynch-pin of mental health.

Dr Lake's Christian commitment was absolute, but various factors militated against a return to the mission field. Among these were his own mission to serve the home church as, over seven years of training and practice, he discovered the extent of emotional problems among the clergy. Christians did not appeal for psychiatric help at that time except in complete desperation. Their own limitations, combined with a lack of proper preparation in the seminaries, meant that many pastors were ill equipped to listen to the deepest concerns of their parishioners.

In his text, *Clinical Theology*, published in 1966, Frank Lake defined his subject as theology rooted in the love and power of God but 'meticulously observant of the sound practice of psychiatry and psychotherapy'. His book was the foundation for the seminars offered in eleven diocesan centres by Lake and his team of pastors and psychiatrists in 1958. The then Bishop of Bradford, Donald Coggan, later Archbishop of Canterbury, had given them his blessing. For many clergymen who attended them they were little short of life changing.

It is not easy to describe, almost half a century later, the sense of liberation and excitement many of those early seminar members experienced, though it has been echoed by many others over the years. Christians striving for perfection – or at least the appearance of perfection - in their lives and character, and inevitably failing, found themselves totally accepted. Frank Lake exposed the over-zealous attempt to win approval from God as less a virtue than a neurotic symptom. He had no use for what he called 'the hardening of the *ought*eries'. Pastors assuming that they had come to learn how to minister to others began to understand, accept and minister to themselves.

What was unique about the Clinical Theology Seminars? Frank Lake had chosen for his model of mental health and normality Jesus Christ himself. The Gospels, notably St John's Gospel, portray Jesus as having a secure sense of identity, rooted in God the Father, the source of his being. When days spent in ministering to the sick, sinful and sorrowful depleted him of resources, he withdrew from the crowds to rest and seek new power through prayer. It was out of this picture that Dr Lake evolved one of the most enduring and flexible of his teaching tools – the Dynamic Cycle.

The Dynamic Cycle charts the process by which a child acquires its sense of being from the love and nourishment provided within its immediate family; then how it is sustained and nurtured so as to be able to overcome the stumbling blocks on the way to adulthood. If all goes well, the adult arrives with a sense of identity and status, in good relationship with others, enabling him or her to lead a productive life and achieve personal goals. However, the stumbling blocks are many. The intricate charts, which accompany the written teaching material, demonstrate graphically how failures at various points in the cycle may affect the developing psyche.

There was also the experiential nature of the seminars, which homed in on real problems and involved every one of the seminar members. The confidentiality of the groups made such sharing possible and bound members together. In addition to these factors, there was Dr Lake's dramatic depiction of personality patterns like the schizoid position, which leads to withdrawal from, rather than attraction to, intimate personal relationships.

From the beginning, the ideas embodied in *Clinical Theology* sparked controversy. At the same time they were warmly welcomed by those who tested them in practice. Frank Lake himself – a slight, modest figure in half-moon spectacles – was an inspiring lecturer and facilitator of groups, much in demand at home and abroad. From his headquarters in Nottingham, he saw individual patients, lectured to seminarians, engaged in tutor-training sessions, and mounted workshops in various aspects of counselling and group facilitation. At annual conferences he introduced new ideas and eminent speakers from abroad, mainly the U.S.A. His outlook was extremely eclectic. The theories of people like Carl Rogers (non-directive counselling), Eric Berne (transactional analysis) and Fritz Perls (gestalt therapy), and many others, became integral to the seminars. It was a fertile period for new approaches to counselling.

Meanwhile, Dr Lake was pursuing his own researches. In his early days in psychiatric practice, he used carefully controlled doses of LSD to assist patients in retrieving past memories and became convinced that the experience of birth was crucial to the defence patterns that people establish in order to ward off mental pain. He soon learned how to retrieve these 'memories' – if that was what they were – even without the use of LSD. And there appeared evidence of pre-birth memories as well, particularly of pain transmitted from the mother. In 1981 he described what he called 'the Maternal-Foetal Distress Syndrome' in his second book, *Tight Corners in Pastoral Counselling*. By the time of his death he was investigating what effect external events might have on the foetus even as far back as its first trimester in the womb.

Frank Lake's restless enthusiasm for new ideas – his own and those of others – did not win him complete acceptance in either religious or psychiatric circles, but he had such devoted adherents in both that the demand for the seminars he had established continued after his death in 1982. The Clinical Theology Association was at first cautious about following its founder into uncharted territory, but has remained faithful to his vision of the counsellor as listener, the mainly silent 'witness to the presence of Christ at the depths of mental pain'.

Under its new name, The Bridge Pastoral Foundation, it offers training seminars, experiential workshops, books and tapes that help to carry on his pioneering work.

APPENDIX 3

1

Has Science so Changed Life As to Make Religious Belief More Difficult?

2

The Nature and Role of the Non-Rational In Human Personality and Life

3

A Scientist's Approach To the Resurrection

HAS SCIENCE SO CHANGED LIFE AS TO MAKE RELIGIOUS BELIEF MORE DIFFICULT?

> 'Down the road someone is practising scales,
> The notes like little fishes vanish with a wink of tails,
> Man's heart expands to tinker with his car
> For this is Sunday morning ……..'

So began a poem by Louis MacNeice written a few decades ago, entitled 'Sunday Morning'. 'Down the road someone is practising scales': would any of us, commissioned to write something typifying Sunday morning nowadays, dream of beginning like that? Of course not, the process of secularization has continued apace, and *we* should have to *begin* with the car.

> 'Down the road someone is washing his car.
> Detergent bubbles vanishing with a pop
> Like good ideas; Man's heart expands
> To tinker with his gears
> For this is motor morning ……'

I am not being fanciful. Bryan Wilson, in 'Religion in Secular Society' says quite categorically: 'The car has been the factor which has altered the pattern of Sunday leisure'.
But of course, one cannot hold the car entirely responsible for the steady process of secularization which has gone on during the 20th Century with ever-increasing momentum. Louis MacNeice's piano has been replaced by a transistor radio tuned to BBC 1 (or Luxembourg), and the weekly hire purchase payments on the piano go instead towards the rental for the telly.
Now, I begin quite deliberately with Louis MacNeice's poem; for although at first sight it might seem that religious practice has little or nothing to do with piano practice, it will be a major part of my contention that the decline in the one has had a great deal in common with the decline in the other. Or I

might have put it the other way round, and said that one might just as well argue that the decline in piano practice is indicative of a decline in the relevance of musical experience to man's life today (or of the non-validity of music as an art form), as to argue that the decline in religious practice is mainly due to increasing evidence of the irrelevance of the religious experience or of the non-validity of religious beliefs in the life of modern man. Indeed, I would go so far as to argue that one can expose some at least of the main reasons for his loss of interest in the life of the spirit simply by analysing his lost interest in piano scales.

It is trite, though by no means trivial, to observe that in the West, at any rate, the affluent society has arrived. Elsewhere, where it hasn't, the basic effort is towards bringing it into being; and far be it from me to denigrate the elimination of poverty and want. I am simply stating a fact, from which a number of other facts stem which I believe are basic to our subject. I believe, in fact, that it is the arrival of the affluent society which is a major – perhaps in the last analysis, *the* major – factor in the loss of concern for the life of the spirit. When Christ said 'It is easier for a camel ....' etc. he was not revealing a hitherto concealed and unsuspected membership rule applicable to prospective candidates for heaven, but stating a simple but profound spiritual truth which seems to hold as inexorably as any so-called law in science. But despite this, we seem increasingly to believe that the wealthier we can get the happier and more contented we shall be, and the nearer will be our approach to heavenly bliss. And so, despite such a programme as a recent 'Wicker's World', on Pools Winners, which showed that almost without exception the big winner finds himself cut off by his very winnings from all real friendship, and that, far from bringing happiness, sudden great wealth almost invariably leads to a state of misery or worse. In spite of such a programme, there is little doubt that the vast majority of viewers who saw it continued their weekly flutter in the hope of achieving the imagined blissful state of monied idleness. In short, the pursuit of affluence is by no means the rational, level-headed, sensible business that its proponents maintain it to be — a worthy successor to what would be regarded as the airy-fairy, pie-in-the-sky of the religious attitude to human life and society. No! it is pursued blindly and irrationally. Bryan Wilson says: 'It is no longer the dogmas of the Christian Church which dictate behaviour, but *quite other* (italics mine) irrational and arbitrary assumptions about life, society, and the laws which govern the physical universe'. In passing, let us note those words 'quite other irrational and arbitrary assumptions', and the subtle and unsubstantiated innuendo against religion implied by them. Spattered all over the writings of the many who are antipathetic to religion, and who write where many will read, is this sort of gratuitous aside, which, threaded through the average layman's reading, comprises a form of brain-washing which he is ill-equipped to resist, or even to detect. I think this insidious process is an important one, continuously at work in the life of the ordinary man today, making belief more difficult by a subtle process of appearing to imply that it is intellectually disreputable. As Gerald Vann has said in 'The Water and the Fire': 'the "way of reason" ..... is not an easy way ..... and if it is the only way available — as often for modern man it is, because of all that he has lost of other aspects of his human heritage — it puts upon the plain man .....a burden which the plain man ought not to be asked to bear. And even if in fact a man *is* equipped to undertake it, still he will have to reckon, as we have seen, with the prevailing climate of opinion, the prevailing scepticism and agnosticism where reason is concerned'.

But to return to the main course of my argument. The blind and irrational pursuit of affluence to which we have just referred is being carried on at an ever-increasing tempo, and the first point I wish to make in relation to science is that affluence is being pursued in this way by means put into men's hands mainly *by* science: and that this, rather than any real undermining of the credibility of the spiritual life, is one of the main ways in which science *is* weakening man's hold on religion

– by offering him a multitude of preoccupying distractions, the spiritual 'fall-out' from which is minimal or non-existent. Science is the hen that lays the golden eggs, and to the ordinary man it appears capable of laying an endless succession of them. That it lays an occasional bad one – like the atom bomb, or the potentially much more evil ones implicit in some modern biological research (to which we might well address ourselves) - is of no consequence to him. As Arthur Koestler says: 'he utilizes the products of science and technology in a purely possessive, exploitive manner without comprehension or feeling' – the very opposite, as Koestler points out, to the art collector's attitude to his treasures, with which he has a participating relationship. As Koestler goes on to say that, by being entirely dependent on science, yet having little or no real understanding of it, modern man leads the life of an 'urban barbarian'. It seems to me that, preoccupied with, and knowing only, the purely utilitarian aspects of science, the layman is divorced from the aesthetic satisfaction which even science might still provide him with, and thus contribute to his spiritual as well as to his material welfare.

So, he has paid someone a fiver to cart away the piano – it took up far too much room anyway – put down the initial payment on the telly, and is becoming increasingly satisfied with living life ever more vicariously from the comfort of his armchair, strategically sited in front of the goggle-box; and for a mere £90 down he can have it in glorious technicolour now. And in a few years from now, as we saw recently in 'Tomorrow's World', the telly will be replaced by the 'World Box', with its revolutionary possibilities for vicarious participation on a global scale. This increasing *vicariousness* of living which modern methods of communication and entertainment – newspapers, radio and television – important as they are, tempt men into, is another aspect I would put high on the list of the ways in which science is lessoning men's first-hand awareness of spiritual things; but we must again note that the process is basically one of distraction from, rather than destruction of, belief.

Its ramifications are legion, and must now be considered. Life's parade passes endlessly across our 17' or 23' screens (according to our means). We get terrifyingly used to seeing people getting shot up or bombed in Vietnam, starving in India, or coshed in Chicago. It is all so near and yet so remote, so much to do with us and yet so little; and human need and suffering is seen to be on such a gigantic scale that men can hardly be blamed for lapsing into paralysed passivity and even cynicism, for only the sensational and horrific is news. Again, as Harford Thomas said recently in 'The Guardian', 'The forces that disrupt our global village are evidently stronger than the communications that bind it together.'

Indeed, the antagonisms between 'Us' and 'Them' seem to have been sharpened by knowing more about each other. To see 'Them' on the television screen is just as likely to excite envy or hatred as brotherly love, and the transistor revolution has made sure that the 'Have Nots' know just what it is that the 'Haves' have. Thus science, innocently enough, stimulates acquisitiveness, and also provides the means by which men are being made to feel puny and increasingly powerless, faced by problems and conflicts beyond their capacity to grasp or to do anything about. (We only have to recall the bewildering barrage of charitable appeals with which we are increasingly bombarded at Christmas time – and under which I personally tend to wilt – to realize the origin of at least some of the apathy around us.) It seems to me that the vastly efficient and rapid methods of modern communications have laid a burden on men under which many have quietly given up, choosing to be little more that spectators from the touchline. In other times the village was the stage on which life was lived, and lived out personally on a scale with which men could still cope, spiritually. I found some support for this notion in Oppenheimer's Reith Lectures of 1953, in which he said: 'Perhaps in the villages the men were not so lonely; perhaps they found in each other a fixed community, a fixed and only slowly-growing store of knowledge – a 'single world'. It is all too often otherwise today, and so

Man tends to live increasingly as Koestler puts it, 'isolated in his artificial environment.' The telly relieves him of the need of, and very often the desire for, all sorts of first-hand living, which he might otherwise have sought. As Gerald Vann says: 'wireless and gramophone are a wonderful consolation for those who cannot be present at a concert, but they are not the equivalent of the concert.' Professionals, from musicians to sportsmen, do things, and can be seen to be doing them from the comfort of the living-room, and doing them much better than we could ever have hoped to, ourselves. And faced with professionalism on the scale and of the quality that we *are* nowadays, little wonder we gave up long ago practising our scales. So passive amusement is displacing creative leisure.

We can have our regular dose of vicarious problem solving, too, in 'Z-cars', 'Softly Softly' and 'Dixon of Dock Green', or of vicarious kitchen-sink experiences in the Wednesday Play and the like; and as Gerald Vann argues, these usually 'bring no catharsis but merely create a craving for daily or even hourly emotional enemas.' Problems of relationship, of belief and doubt, and of making difficult decisions we leave increasingly to the people on the other side of the screen, and to our relief we find that there are more and more people only too eager to tell us what we can believe, and what we can no longer go on believing, if we are to retain our self-respect among those who are 'with it'. Our minds are increasingly being made up *for* us by the expert makers-up of minds, the subtle and insidious brain-washers, and with their best bedside manner. These include not only the advertisers on I.T.V. who steadily whet our consumer appetites, but also those who in the brief span of a quarter-hour's talk on the Third Programme may give us, on the one hand the 'low-down' on the significance to the nuclear physicist of the existence of a magnetic monopole, and on the other an explanation of the so-called 'death of God', or a thumb-nail sketch of the Essenes as drug-takers, or better still treat Dr Leach's theme that men are like gods, thus 'buttoning up' for the poor layman what might have cost him half a lifetime of essential anguish and doubt - anguish and doubt that might have spelled his ultimate salvation.

These potted programmes give him the comforting illusion of being 'genned up', and if he supplements them with a little reading it will probably be to the 'Readers' Digest' that he turns. In particular, programmes and mini-articles on the achievements of science by the Chapman Pinchers of this world leave him in no doubt that in science the difficult can be achieved immediately and the impossible takes only a few months longer. Needless to say he is dazzled by the prospects. And so – heavens above! – why on earth should he concern himself with God or the Church – the golden hen clucks every few seconds in his service in one way or another. The Church no longer appears to offer him anything he really - deep down - feels the need of. Admittedly he will often seek to be married by the Church, have his children baptized by the Church, and his kith and kin buried by the Church, but increasingly, in these matters, he is simply being carried along on a current of belief which still flows in from the past, but to which he himself imparts no further momentum. Am I exaggerating? I think not. If I may quote Bryan Wilson again: 'For many the Church functions as a service agency, providing appropriate ceremonial for prestige and status-making at crucial stages in the life cycle.' And again: 'The Church plays its part in the lives of many more as a service facility than as an evangelistic agency, as the provider of occasional and reassuring ritual than as the disseminator of vital knowledge or the exemplar of moral wisdom.' Or yet again: 'The Church appears increasingly as some department of the welfare state to be used as and when the individual requires its services.'

Bryan Wilson excels himself when, having said that the majority of people still seek the services of the Church in marriage and that a still greater majority are involved with the Church at death, he adds with grim humour: 'Indeed, a man needs extraordinary presence of mind at death if he is to avoid religion officiating at his burial'... But even these remaining functions of the Church, at marriage, baptism and burial, are seen by most sociologists to be sought in the last analysis for mainly superstitious reasons. For the rest, modern man lives in an entirely pragmatic environment, and he feels that the Church 'gets you nowhere'. As Bryan Wilson says yet again: 'Pragmatic values are manifestly more implicit in the scientific enterprise than in religion'; and he goes on to point out that whilst it is true that in many ways the Church has tried to adjust to the new situation by endeavouring to make its own approach more pragmatic, nevertheless he contends that the real danger of science to religion is in the increased prestige of science and the decline in the intellectual prestige of religion. 'Since science had answers, and had positive and tangible fruits, it came increasingly to command respect and approval'.

In short, science works, and can be *seen* to work. Indeed, it answers the question 'how' so well and so often that I believe that men have become almost completely preoccupied with the 'how' questions, bothering to ask 'why' less and less as time goes on. In fact I do not think it would be exaggerating to say that science has answered the 'how' question so often now, and so well, that the layman is befuddled, and with his lines crossed, has come to believe in the back of his mind that science answers the 'whys' as well as the 'hows', and that it does so infinitely better than the Church ever did. He has come, in fact to believe – quite erroneously of course, and that is the tragedy of it – that science knows *all* the answers.

Needless to say, we all know that the Church has contributed its full share in the past to this confusion, and served its own cause ill, by usurping what is the proper role of science, claiming to be able to answer the 'how' questions as well as the 'why' questions, which are its own true territory. Consequently, as Whitehead has said: 'the theologians of the 16th and 17th centuries were put into a most unfortunate state of mind. They were always attacking and defending. They pictured themselves as the garrison of a fort surrounded by hostile forces. As a result religion found itself on the defensive, and on a weak defensive, over the last two centuries.' Something which had been claimed to be vital *has*, as Whitehead says, 'finally, after struggle, distress and anathema, been modified and otherwise interpreted. The next generation of religious apologists then congratulates the religious world on the deeper insight which has been gained. The result of the continued repetition of this undignified retreat during many generations, has at last almost entirely destroyed the intellectual authority of religious thinkers.'

It has indeed been a sorry story, from Galileo to Darwin and beyond, not helped, either, by schizophrenic tendencies from time to time within the Church, such as is exemplified by the story of the priest who, confronted with a fossil specimen by a young friend with the comment that it must be 'as old as Creation', is said to have replied, gravely shocked, with 'Oh no! *Much* older than that!' Little wonder that the befuddled layman has concluded that the Church is no place for him if he wishes to claim even a *nodding* acquaintance with the intellectuals.

The point I wish to reiterate strongly is that he is neither sufficiently informed about science or religion to have any perspective of his own in the matter, so that he is at the mercy of the most vociferous of the professional persuaders, in what little time he has left over from getting and spending with the means put into his hands by science. To summarize, so far as the layman is concerned, I would suggest that science has not in actual fact undermined his religion by making it less credible – although he may *think* that is what science has done. It would be truer to say that it has innocently enough weaned him *away* from religion by so changing the climate and material circumstances of his everyday life that he has come to feel that there is little or no role left for religion to play in his life: in a word he's not interested because quite literally *he sees no use in it*. And because science is ultimately largely responsible – though without malice aforethought – for bringing about this state of affairs, we can, *in this sense,* argue that science *has* changed life so much that men *in fact* do look at it less and less in a religious way, although science has *by no means* destroyed the credibility of religion. And this weaning-away process, let it be noted, has involved not merely the lure of material pleasures and treasures which science dangles in ever-increasing variety before men's eyes; it has involved an even more insidious destruction of 'silence without, and quietness within.' The ordinary man would deride the idea of being still in order to know. As Gerald Vann has said: 'All day long the radio blares, all the time; whatever he is doing there must be 'background music'. The radio is a conduit through which pre-fabricated din can flow into our homes and lives'. Knowledge for modern man (and technological science must be held responsible to a large extent for this conditioning) is, says Gerald Vann, 'more and more exclusively a utilitarian scramble to acquire knowledge of facts and especially material facts, and above all, commercially rewarding facts. Wisdom must give way to efficiency, culture to commerce .... the vast triumphs of science have caused us to neglect and perhaps deride the other avenues of knowledge ..... sense, feeling, intuition'.

In a word, modern man, I would contend, is dangerously lopsided, and if I may extend the metaphor, in danger of capsizing, with his head far too big for his body, as Kapek portrayed the Inventor in the Insect Play.

Before moving on to my last topic I would therefore put before you what I feel to be a special responsibility laid upon such as ourselves who are professional scientists and nevertheless have some awareness of the needs of the layman, lost – and hardly aware that he *is* lost – in the no man's land across which science and religion appear to face each other with hostility from prepared positions, and with little or no chance of reconciliation. That responsibility as I see it is one of interpretation and mediation, until our fellows become aware at last – even though it be dimly – that the apparent clash between science and religion is in fact in Whitehead's words 'a sign that there are wider truths and finer perspectives within which a reconciliation of a deeper religion and a more subtle science will be found.' This is no mean responsibility, and we cannot argue that our opportunities for discharging it are scant, with the University virtually on the Cathedral's doorstep. [As is the case in Guildford.]

Finally, then, let us turn from the layman in his no man's land, to the militantly atheistic professional scientist in his laboratory, and try to reach a better understanding of his psychology. It might be helpful for a moment to consider some of the traditional attributes of the scientist: curiosity, coupled with a clue-cracking complex. In other times we would need to have added a sense of mystery and wonder, but alas, most scientists these days, from the humblest to the greatest, would blush to have this attributed to them. Yet Einstein is reported to have said that 'Whoever is devoid of the capacity to wonder, whoever remains unmoved, whoever cannot contemplate or know the deep shudder of the soul in enchantment, might just as well be dead, for he has already closed his eyes upon life ….'. According to Koestler, 'the feeling is the common source of religious mysteries, or pure science, and art for art's sake – their common denominator, and emotional bond.' Gerald Vann says: 'We must be contemplative before we can be wisely and graciously active': he goes on to say: 'The great scientists knew this, the great philosophers knew it, the poets, the artists and makers of music knew it, but we have lost the sense of wonder'. One can well imagine the impatience of Dr Leach (whom Lord Soper good-humouredly accused in a recent discussion of shedding more heat than light on the subject) if he were faced with Whitehead's definition of religion: 'Religion is the vision of something which stands beyond, behind, and within, the passing flux of immediate things, something which is real, and yet waiting to be realised; something which is a remote possibility, and yet the greatest of present facts; something that gives meaning to all that passes, and yet eludes apprehension, something whose possession is the final good, and yet is beyond all reach; something which is the ultimate ideal, and the hopeless quest.' 'Eluding apprehension' …. 'beyond all reach' ….

'the hopeless quest'. If with Dr Leach you believe that men have become like gods, such phrases are the raw material of apoplexy. But, as I have said elsewhere, I believe with Thomas Merton that if we are to be truly filled, we must first be prepared to be 'stripped and poor and naked within our own souls'. There must be the 'self-naughting' of the mystics, the 'going out from all things', the 'total self-dispossession', before we can be filled.

Koestler argues that the good scientist is a nice balance between self-assertive and self-transcending elements. It would seem that the self-assertive element has in these days come very much to the top.

But let us look for the moment at the reasons — some hidden — that may make a man choose science as a career. I was in a restaurant last week sitting at the same table as a young man discussing with an older man his choice of chemistry as a career, for which his training had just begun. He remarked that he was having second thoughts about his choice and his motives for it. The older man, who from the way he spoke was himself a scientist, said something which ran roughly as follows: 'Well, let's face it. Those of us who choose something like physics or chemistry do so in the first place as a form of self-indulgence. We like messing about in a lab. with this and that, and think it would be nice to spend our lives messing about in the same sort of way'. Self-indulgence may seem a rather strong word for it — but I wonder. Such self-indulgence naturally leads to the self-assertion to which I have just referred: problem-solving, with the prospect of publication to follow, and the kudos that it will bring, at any rate among the relatively small elite of the initiate; or if it happens to be a golden egg, then — who knows — fame and fortune overnight, perhaps.

But I still don't think this is the whole story. I believe that there is a factor at work in the choice which is not often talked about, and would be less willingly admitted to, but which is probably present to some extent in the make-up of the majority of those choosing science as a career, particularly one of the non-life sciences. Inanimate matter is easier to deal with in the last analysis, even though the analysis may involve the 'higher mathematics', as Chesterton called it, and in which, according to him, 'the whole of the house is upside down'. And I have more than a suspicion that those of us who choose a career in a non-life science have an innate sneaking preference for that kind of topsy-turviness, compared with the topsy-turviness and unpredictability which one faces in the life sciences, or a job concerned largely with human relationships, and the problems which go with them. As I saw it stated recently; 'The scientist cuts his facts to his own emotional limitations.' Here I see eye to eye with Dr Leach for a change. In his first Reith Lecture last autumn he said: 'In the world of science different levels of esteem are accorded to different kinds of specialist. Mathematicians have always been eminently respectable, and so are those who deal with hard lifeless theories about what constitutes the physical world: the astronomers, the physicists, the theoretical chemists. But the more closely the scientist interests himself in matters which are of direct human relevance, the lower his social status. The real scum of the scientific world are engineers and the sociologists and the psychologists. Indeed, if a psychologist wants to rate as a scientist he must study rats, not human beings. In zoology the same rules apply. It is much more respectable to dissect muscle tissues in a laboratory than to observe the behaviour of a living animal in its natural habitat. If you enquire from the scientists themselves as to why they have these valuations you will find that it is the regularity and order of the physical sciences which are admired. The biological sciences come to be respected precisely in the degree to which they can make exact predictions. Conversely, the social sciences and the practical men are condemned because they are imprecise, and because they are "not sufficiently detached". The underlying psychology here is complicated. The scientists are engaged in exploring a changing universe but they are frightened, just like the rest of us, by the idea of a changing society. So they try to keep scientific activity and social activity apart and pour contempt on those who get them muddled up. Good science is 'pure' science, and must on no account be contaminated with real life.' So much for Dr Leach for the moment.

What it amounts to is that the non-life sciences can be, and often are, a kind of escape into what Frank Lake, the founder of the Clinical Theology movement, calls the 'ivory tower of intellectualism – a schizoid tendency to shy away from real-life commitment in relationship.' Of course this is a generalization, but if it offends, if you feel yourself bristling at the very idea, I would ask you to examine the nature and roots of its apparent unpalatability. May it not be that there is more than a grain of truth in it? There certainly is for myself, despite the fact that I have spent more than twenty years of my working life in daily contact with cancer patients.

So: our typical scientist may well be someone whose predispositions have already orientated him away from the affairs of *men* and towards inanimate *matter* in one form or another. The self-assertive side of him gains its fulfilment in problem solving, and in the kudos gained from the subsequent publication of his solutions, and much more rarely, the fame and fortune which may come his way if he hits the jackpot. As often as not, preparation for such work involves *training* (rather than education) which as Dr Joseph Pieper said is concerned 'with one side or aspect of man with regard to some special subject.' He contrasts training with education, which, as he says, concerns the whole man. In fact, as I have pondered this evening's topic, I have come to feel that, one of the dangers of our time is that we tend to be technologists rather than scientists in the fullest sense if the word.

Gerald Vann says: 'the personality as a whole is no longer engaged'. I believe that this is the reason why, to a man like Allegro, tickled pink with his recent hypothesis that the Essenes were drug addicts, it doesn't even occur to ask how a body of teaching and a quality of life such as is enshrined in the Four Gospels could be a mere by-product of the psychedelic fantasies of the latest recruit to a group whose way of life was to go 'high' regularly. Here a schizoid tendency to let reason take over at the cost of everything else has become almost schizophrenic: Koestler's self-transcending element has become wholly submerged under the self-assertive, and megalomania looms over the horizon.' These are strong words I know, but the temptations are enormous, and the material successes of the physical sciences serve only to magnify them. For Dr Leach 'Men have become like gods' …. 'Science offers us total mastery' … 'All of us need to understand that God, or Nature, or Chance, or Evolution, or the Course of History, or whatever you like to call it cannot be trusted any more. We simply must take charge of our own fate'. 'Only those who hold the past in complete contempt are ever likely to see visions of the new Jerusalem'. 'It is time that we had done with the idea that humility is a virtue …. the young have got to learn that *they* are masters of the situation'.   And so on .. Are not these strong words, too?

I believe that, more than any factual and material evidence that a scientist may discover in the course of his work which might lessen the credibility of religion for him, it is this loss of humility, this loss of the self-transcendent element of his personality, this loss of a sense of wonder, which turns the potential scientific atheist into an actual one. As Gerald Vann says: 'intellect, scientific, analytical, practical, has been abused, has been developed and idolized at the expense of the psyche as a whole; if we are to return to sanity it must be through the return of intellect to its fundamental purposes, to wonder, adoration, vision, wisdom, all of which can spring only from humility and inwardness.'

But Dr Leach, at any rate, does not seem to have reached the point of no return. Near the end of his sixth and last lecture he discusses the early cave paintings of wild animals at Lascaux, 'tamed', as he puts it, 'by the magic ritual of paint thousands of years before they were tamed domestically'. He goes on: 'Art and poetry are the power to transform, the ability to take nature to pieces and recreate it: it is dangerous but it is magical, and it has been man's heritage from the beginning ... If you want to find a way out of our modern dilemma, you should talk with artists and poets rather than with University dons.' Is that not a surprising conclusion to have reached from his earlier premises?

It was otherwise with Darwin. He was what Koestler calls a 'one-idea man' and spent his lifetime working on it. Koestler points out that he lacked the many-sidedness of men such as Newton and Maxwell and, after evolving his theory, his life could be described as one of 'duty, devotion to task, rigorous self-discipline and spiritual desiccation', apparently. Thus, before he evolved his theory of evolution, deep in the Brazilian forest he had felt a quasi-mystical 'deep inward experience' as he called it, that there must be more in man than 'the mere breath of his body'. After his theory had been produced, he experienced, in his own words, 'a curious lamentable loss of the higher aesthetic tastes.'

An attempt to read Shakespeare bored him (as he reported) 'to the point of physical nausea'. Later, in his autobiography, he complained: 'For many years I cannot endure to read a line of poetry. My mind seems to have become a kind of machine for grinding general laws out of a large collection of facts, but why this should have caused the atrophy of that part of the brain on which the higher tastes depend, I cannot conceive. The loss of these tastes is a loss of happiness, and may possibly be injurious to the intellect, and more probably to the moral character, by enfeebling the emotional part of our nature.'

Gerald Vann says: 'Holiness is the fruit of love and wisdom: and the first essential step to these is the sense of wonder.' I submit that our age is a technological one rather than a truly scientific one, and that love and wisdom and a sense of wonder have little or nothing to do with technology as such .. Wonder is closely linked with worship in the human heart. Whitehead says: 'The immediate reaction of human nature to the religious vision is worship.' He goes on: 'Worship is a surrender to the claim for assimilation, urged with the motive force of mutual love'. 'Surrender', and 'assimilation': Thomas Merton speaks of becoming 'like vessels that have been emptied of water that they may be filled with wine', and of our faculties being 'emptied of their desire and tension towards created things'. I suggest that this includes the scientists' desire and tension towards the raw material of his particular chosen discipline.

As I wrote those words I had a sort of vision of Christ, face-to-face with a present-day scientist who asks him 'What lack I yet?' Christ answers: 'If thou wilt be perfect [or 'If thou wilt be *mature*' as I have seen it suggested as being nearer Christ's meaning] go, divest yourself of all your sense of intellectual wealth and power, put aside your enchanting problems for a while, and enter into the experience of those who are poor and stricken in spirit ..... and come and follow me'. But when the scientist heard that saying, he went away sorrowful, for he had great intellectual possessions.

I have tried to keep to my brief: 'has science made religious belief more difficult?'. Apart from touching lightly on the Church's disastrous sallies into the 'how' world, I have not dealt at all with the Church's contributions to the difficulties. Alas, they are legion – but that is another story.

REFERENCES AND BIBLIOGRAPHY

1. Barnes, Kenneth C: 'The Creative Imagination', Swarthmore Lecture 1960, of the Society of Friends. George Allen and Unwin.
2. Cotgrove, Stephen: 'The Science of Society', George Allen and Unwin.
3. Koestler, Arthur : 'The Act of Creation', Hutchinson.
4. Lake, Frank: 'Clinical Theology', Darton, Longman and Todd.
5. Leach, Edmund: 'The Listener', 16th Nov. 1967 to 21st Dec. 1967
6. Merton, Thomas: 'New Seeds of Contemplation'; also 'No Man is an Island', Burns, Oates.
7. Vann, Gerald: 'The Water and the Fire'; also 'The Divine Pity', Fontana Paperbacks.
8. Whitehead A.N. 'Science and the Modern World', Cambridge University Press.
9. Wilson, Bryan 'Religion in a Secular Society', Watts.

THE NATURE AND ROLE OF THE NON-RATIONAL IN HUMAN PERSONALITY AND HUMAN LIFE

Kierkegaard said 'Lying is a science, truth is a paradox'. Commenting on this, Frank Lake, the pioneer of Clinical Theology, said, 'Certainly the truth about the human heart and mind of man is that it is 'deceitful above all things'.' Frank Lake was referring, of course, not to the deception of others, but to a certain kind of *self*-deception which goes on all the time regarding our own inner world, and without which, in fact, many, and perhaps the majority of us, would sooner or later break down. For he adds at once 'Things (inside us, that is) are seldom what they seem to be. To reduce this complexity by a superficial science of behaviour, to observable morality, is to embrace a lie'. Again, Guntrip has said 'Mental processes in a human being are 'personal processes', and it does not appear relevant to abstract them in such a way that they are treated as impersonal processes'. Yet again, Jung, in his little book 'The Undiscovered Self' said 'The individual is not to be understood as a recurrent unit, but as something unique and singular which in the last analysis can neither be known nor compared with anything else', and he goes on to say 'If I want to *understand* an individual human being' (and he distinguishes between '*understanding*' and '*having knowledge of* ') 'I must lay aside all scientific knowledge of the *average* man in order to adopt a completely unprejudiced attitude'. The reason Jung gives for this necessity is that 'scientific' education is based in the main on *statistical* truths and *abstract* knowledge and therefore imparts an unrealistic [note this] 'an *unrealistic*, rational picture of the world, in which the individual, as a merely marginal phenomenon, plays no role.' And he adds 'The individual, however, *as an irrational datum*, is the true and authentic carrier of reality, the *concrete* man as opposed to the unreal ideal or 'normal' man to whom the scientific statements refer'.

Repression is almost a dirty word in a permissive society, yet it is a fact that for most of us the truth about ourselves –

about many, perhaps most, of those factors which have been ultimately responsible for shaping *our* individuality – would be unbearable, so they are repressed to the deep unconscious early in our development; and it is the attempt to describe the complexity of the resultant personality structure in any sort of superficial or general terms which Frank Lake was asserting would be tantamount to a lie. Thus self-deceit of the kind Frank Lake, and indeed, any psychoanalyst is concerned with, is, in fact, a very important mechanism at work in all of us, whether or not we care to admit it.

The role of this particular form of self-deceit is by no means a purely negative one either, confined merely to the maladjusted or neurotic. Thus Alex Comfort, in his book 'Nature and Human Nature' - from which I shall, in fact, quote extensively, using it as a kind of anvil on which to beat out some of my own thoughts – says: 'It is not impossible that ..... repression and a mind divided actively into conscious and unconscious levels ...... produced the most significant adaptation in mammalian history, the emergence of conceptual thought. .... The need to repress so much early experiences into the (aboriginal) 'dream-time' may have forced the consciousness to think in the straight-line characteristics of human conceptual thought'. .... 'There is the effect' he says, 'of unconscious forces, not only in making us unable to see the obvious, which is a handicap, but in pre-disposing us to see the unobvious, which may make us geniuses'. Comfort, having speculated that both Darwin and Freud 'probably owed their unusual insights to peculiarities of personality which arose in their infancy', goes on to say: 'It appears that the human mind, unwieldy and unreliable as it may appear when we consider the unconscious forces at work on it, is actually a more valuable matrix of ideas and inspiration precisely for that reason – and, for *pure* reason, would be *un*motivated, save by idle curiosity'.

In spite of all this, the frank admission to ourselves that our personalities are by no means identical with our 'persona' is almost inevitably felt as a threat: Jung says 'It seems a positive menace to the ego that its monarchy can be doubted'. From this threat we all have at least one or more 'ivory towers', as we may call them, into which we retreat at the first sign of trouble. Indeed, we may be dwelling more or less permanently in one of these embattled residences; and, thank you very much, we are very comfortable; we don't much care for the view from the window, and as for ascending to the battlements, or worse still, rummaging around the contents of the altars dumped there years ago either by ourselves or our forbears – well – do it if you must, but don't expect *me* to accompany you! This identification of our psychic lives – indeed, of our very selves – with our conscious personalities, and thus with 'the straight lines of conceptual thought' that go with it, is far from innocuous; for, as Jung says, since it tends to be universally believed that 'man is merely what his consciousness knows of itself, he regards himself as harmless, and so adds stupidity to iniquity'.

Perhaps to those that come after us, it will seem in this sense that the most impressive Ivory Tower of all erected in our day and age was natural science itself – especially non-life science. I am sorry if that sounds a little harsh or traitorous, but introducing the passage from Frank Lake which I quoted at the beginning was the following: 'The scientist in secret despair of becoming a person, posits his own being in terms of rationality, *neatly omitting the existential question.*' (Italics mine). 'He offers as proof of himself a few impressive but really unimportant certainties' (unimportant, that is, existentially). Cogito, ergo sum. Someone else has said 'The scientist cuts his facts to suit his own emotional limitations', whilst Dr Leach, in his Reith Lectures two years ago said: 'In the world of science those who deal with hard lifeless theories ..... have always been eminently respectable ..... the more closely the scientist interests himself in matters which are of direct human relevance, the lower his social status .... if a psychologist wants to rate as a scientist he must study rats not human beings .... If you enquire from the scientists themselves as to why they have these valuations, you will find that it is the regularity and order of the physical sciences which are admired …. Conversely, the social sciences and the practical men are condemned because they are imprecise, and because they are 'not sufficiently detached' ….. Good science is pure science, and must on no account be contaminated with life'.   So much for Dr Leach. Alex Comfort, reflecting on scientists in relation to artists says 'They (that is, the scientists) are equally drawing on, driven, and inspired, and handicapped by, the emotional resources in Man'. But he adds dramatically, *'As a group, they are commonly refugees from these resources, however ….'* (Italics mine).

What a dressing-down all this amounts to! And I venture to suggest the reaction of most of the scientists present at the lecture was at least *tinged* with indignation. I know mine was at first. And if we *did* feel a little maligned and misunderstood, then that fact itself is noteworthy; perhaps we might have reacted a little more appropriately with a simple *'touché'.*

We are concerned this evening with what we may call the 'irreducible non-rational element in human personality and life, and so far I have endeavoured to remind us that our *conscious* selves – that part of us which Alex Comfort says has learned to think 'in the straight lines of conceptual thought' – is, as we know, far from being the *whole* of us, and that it is, in fact, even *motivated* by the hidden, non-rational depths of our beings, in the scientifically orientated and non-scientifically orientated, alike. And all of us, are, to a greater or lesser extent 'refugees' – to use Alex Comfort's word again – from these hidden depths within us, refusing, as it were, to grant full conscious recognition to their existence, fearful lest in so doing we should enable them to become a break-away state, so to speak, and over-run our conscious selves with which we all tend to identify, for most of our waking moments; it is a different matter, of course, when we are asleep.

At this point it is, perhaps, high time we defined our terminology a little more carefully. For myself, I have preferred the word '*non-*rational', rather than 'irrational', indicating the *ob-*verse rather than the *re-*verse of 'rational'. The dictionary defines 'rationalism' as 'treating reason as the ultimate authority' and '*ir*rational' as '*un*reasonable, illogical'. '*Ir*rational' is, in fact, a somewhat coloured – even pejorative - word, often used to describe behaviour which is distinctly odd, bordering on the bizarre, or sufficiently abnormal as to suggest the presence of mental illness. For that reason, in referring to that irreducible 'otherness' in us, distinct from our thinking, rationalizing, intellectualizing selves, I prefer to use the word 'non-rational'.

We shall now turn our attention then, to consider rather more systematically the nature and role of the rational and non-rational elements in our personalities. In this matter Alex Comfort seems frequently to find himself on the horns of a dilemma, for whilst continually having to admit the enormous influence of our non-rational depths on our conscious attitudes and actions, he has obvious difficulties in summoning any appreciable degree of respect for the well-springs of our actions. 'The human situation', he says, 'must be unique, in which, as it were, directives and messages come in to the Head Office from a clandestine department that the staff do not realize is there, with its own logic, its own aims, *and the solidified inner attitudes of a child of three*' (italics mine). Yet he goes straight on to add that 'if we *did* manage to get rid of our unconscious processes …. we should become decerebrate – deprived of a working brain and of all desire, motivation, inspiration, or mental activity recognizably human, for (he adds) '*it is our unbiddable half which is the powerhouse for all such matters*'. I'm afraid my only comment on that must be 'Some responsibility for a child of three!' Whilst fully recognising the infantile nature of much of the material buried far down in our non-rational depths, 'it stands to reason', as we might say, that there must be more to those depths than that. Jung certainly thought so when he said that the unconscious is 'the only accessible source of religious experience …. the medium from which religious experience seems to flow'. Alex Comfort, however, seems to see the contents of our unconscious as largely problem children – filled albeit with restless energy –

'feelings, irrational fears, and anxieties' which we have got to learn to manage '*without denying either them or our reason*'. But alas! Comfort's reasonableness forsakes him a few pages on, when, evidently labouring under what C H Dodds once referred to as 'the scandal of particularity' which is at the heart of Christianity, he refers to religion which has become 'contaminated with historicity' as dealing with what he quaintly calls 'untrue facts' instead of confining itself to feelings. For him this represents 'going into reverse, of cultivating, not exercising, irrationality – of denying the real in the pursuit of the wish'. We might well ask at this point, reframing Pilate's famous question slightly, 'What is reality?' Leslie Dewart, in 'The Foundation of Belief' seems apparently to hint that the concept that *being* is to be equated to reality may need to be transcended. This is a somewhat tenuous line of thought for me, and I found myself much more at home with John Robinson when he said, in 'Exploration into God', that in future 'the reality may have to be witnessed to much more indirectly, obliquely, parabolically, brokenly – in action, in suffering, rather than in words'. This is the language of our *non-*rational selves.

Alex Comfort's juxtaposition of 'the real' as over and against 'the wish' cannot sustain its implications, for a 'wish' is simply a 'desire', and by no means all the things we wish for or desire are creations of infantile fantasies, and there *are* times when it is certainly pertinent to ask 'What is the reality behind the wish – the half-apprehended truth which evokes in us the desire to come to closer terms with it?' Comfort obviously does not believe that the wish ever subsists on a reality, for he speaks of the 'fundamental quality of science' as being 'the acceptance of truth over the wish'. Here we might well return to the original form of Pilate's question and ask 'What is truth?' – with the other question which it inevitably implies: 'And how do we recognize it when confronted by it?' And *is* the only kind of truth worth possessing that which we, as scientists deal with – namely that based on observations verifiable again and again, indefinitely, given only the same set of conditions? *Is* the only kind of knowledge we really value that which is based on rational thought processes? I believe that more and more of us in these days are coming to hold albeit a sneaking belief that this is so, discounting the non-rational side of ourselves, with its own special needs and logic.

The reason for this is, I am sure, to be found quite simply in the conditioning to which we have all been subjected, as a result of the truly fantastic progress of the material sciences in the last hundred years or so. 'Scientific knowledge', says Jung, 'not only enjoys universal esteem but, in the eyes of modern man, counts as the only intellectual and spiritual authority'. A hyperbole, perhaps, but as I myself said here in a talk two years ago – if I may be forgiven a personal quote – 'Science is the hen that lays the golden eggs, and to the ordinary man it appears capable of laying an endless succession of them. That it lays an occasional bad one (like the atomic bomb) is of no consequence to him …. In short, science works, and can be seen to work. It answers the question '*How?*' so well and so often that men bother to ask '*Why?*' less and less. In fact, the befuddled layman has come to believe in the back of his mind that science answers both the 'whys' as well as the 'hows'. He has come to believe – quite erroneously, of course – that science knows *all* the answers, and that knowledge which cannot (from its very nature) be tested and verified 'scientifically' is at least suspect, if not worse'. Do not misunderstand me. I am not for one moment suggesting that we should do *violence* to our intellects. What I *am* suggesting, however, and I make no apologies for doing so, is that it is not only the befuddled layman who finds himself in this position, but that we ourselves are in danger of failing to be really honest with ourselves about the attitude and temper of our own minds in this matter. Not many of us in these days see as clearly as did Simone Weil, the nature of what she called the 'intellectual vocation', nor are we nearly as aware as she was of its natural and inevitable limitations. 'My vocation', she said, 'imposes on me the necessity of remaining outside the church…. for as long as I am not quite *in*-capable of intellectual work. And that is in order that I may serve God and the Christian faith *in the realm of the intelligence*' (and note the clear implication here that there are other realms of service).   A little later on she speaks of this vocation as *depriving* her of 'sharing in (Christ's) flesh in the way he instituted'. The point bears emphasizing that Simone Weil did *not* imply - as so many do in these days either explicitly or implicitly including many of our theologians who make the headlines – that what she called 'the intellectual vocation' is the only really valid and respectable one left to us, and that intuition, for example, is the dubious and indeed dangerous faculty manifested by women when driving cars... I might add in parenthesis that although Simone Weil's 'intellectual vocation' in her own words demanded that her thoughts 'should be indifferent to all ideas without exception, including for instance materialism and atheism, and equally welcoming and equally reserved with regard to every one of them', she nevertheless affirmed elsewhere that Christ had on *one particular* occasion (how offensive 'particularity' tends to be to the scientific mind!) simply 'taken possession' of her, and that in this sudden possession 'neither (her) senses nor (her) imagination had any part'. 'I only *felt*,' she said, 'the presence of a love, like that which we can read in the smile of a beloved face'. This is, of course, *mysticism*, and it is bound to evoke an impatient cough and a shuffle of feet among the rationalists. But must 'mysticism' join 'repression' as a dirty word in the context of our present-day neutral climate?

Why *are* we in these days so often 'refugees' from our deeper selves? Why *do* we place such emphasis on what can be 'measured and weighed' by the scientist and with increasing impatience dismiss the 'imponderables' of life.   Jung speaks of the state of 'uprootedness' which our 'concern with consciousness at the expense of the unconscious' brings about, causing us to slip imperceptibility, as he says 'into a purely conceptual world *where the products of (*our) *conscious activity progressively replace reality.*' (italics mine). Charles Raven once said that 'the intellect is the *interpreter* and not the *originator* of experience'. Are we sure we haven't got it the other way round? Is modern man, in fact, stalking about his world like the Inventor in Kapek's 'The Inspect Play' – top-heavy with a head that threatens to over-balance him? Is it not high time that we began to redress the balance a little? Jung warns of the 'bitter revenge' that our *non*-rational depths will otherwise take on us. As Damaris Parker-Rhodes, a well-known Quaker, has said: 'The mystery of eternity in time has to be inwardly received through the process of living, and *cannot be received through the mind, just as information*' (italics again mine).

How *do* we receive the mystery? We should indeed be *sub*-human if we thought it was with our intellects alone. Philip Martin, in the Introduction to a wonderful little book called 'Mastery and Mercy', which is, in fact, a commentary on two poems, 'The Wreck of the Deutschland' by Gerard Manley Hopkins and 'Ash Wednesday' by T S Eliot, says: 'For many years Christian teachers have been content to communicate Christian truth mainly in rational terms, and to present the Gospel largely by way of intellectual concepts. The result has been that, only too often, men are not *moved* by what they (the teachers, that is) profess to believe…. True religion must touch and move us, in much the same way as great poetry touches and moves us. For in poetry truth is communicated not so much rationally through the mind as obliquely through the imagination; and the subsequent reader is led to take the truth into himself rather than having it presented (merely) to his mind' . Martin goes on: 'The words the poet uses to express truth create an emotional impact. They work upon the imagination of the reader in such a way that the truth is communicated at a deeper level than that of the mind or the emotions'. According to Martin, Hopkins himself said that there is a kind of clearness in poetry by which 'meaning, if dark at first reading' – when once it is discovered 'will explode'. Martin comments: 'Such an "explosion out of darkness", sudden revelation of meaning, is always a most satisfying, and salutary, experience; in this poem ('The Wreck of the Deutschland') the truths are the fundamental truths of the Christian religion, and precisely by reason of a preliminary "darkness", "explosion", when it comes, bursts at a deep level in the reader; both the mind and the will have applied themselves to understanding, and, as a result, the truth is apprehended deep in the imagination and the soul'. I would like to discuss this whole passage in detail, but must content myself with merely asking you to note the use of the word *apprehended* rather than 'comprehended', and to contend that although there is a difference between the two, it is in the *kind* rather than the *quality* of knowing that the difference lies.

Of course, if we are subscribers to 'the brain as computer' concept, this whole matter of 'insight', this 'explosion out of darkness', can be described in very different terms: up to a point 'Yer pays yer money and yer takes yer choice'. Thus Alex Comfort speaks of 'insight' which 'is not information so much as feeling'. Having warned us of the dangers of admixing 'this type of religious (sic) experience' with the 'dogmatic religions' which, he adds 'can produce our unjustified sense of conviction, proof against reasonable argument', he goes on: 'On the other hand, taken as simple devices to generate states of feelings they (the first type of 'religious experience') could prove extremely useful'. Comfort is willing to admit that the experience of listening to a great symphony is of 'an oceanic sensation', but having said that, he qualifies for the Mr Bathos of All Time, when he adds: 'We do not underrate the beneficial effects because [the symphony] imparts [the 'feeling' kind] of insight without imparting information'. (Such experiences) 'interest us now because it looks as if the human brain is programmed to benefit from the 'oceanic' sensation which these various devices (sic), suggestive and physiological, produce. No doubt the sensation itself is a cerebral trick, a final common truth which can be set off by various means ....' So there you have it! - the 9th Symphony amounts after all simply to the astute exploitation by Beethoven, of a mere cerebral trick which our brains play on us.

Well, as I say, 'Yer pays yer money ….', but for my part I think I can recognise bathos when I see it.

But to return for a moment to Martin who is much more to my way of thinking. 'In poetry', he said, 'truth is communicated, not so much *rationally* through the mind, as *obliquely* through the imagination.' I could not help being reminded at this point, of a situation which arose in atomic physics forty years or so ago, which is by way of providing us with a useful simile, if nothing more. The nuclei of atoms are positively charged, and if, by means of a so-called particle accelerator, we fire positively-charged particles such as protons at atomic nuclei, the positive charges on the latter repel the positively-charged particles fired at them, slowing them down, bringing them to a standstill, and driving them away again, *unless,* that is, the bombarding particles have enough energy to surmount the so-called *potential barrier* which surrounds each atomic nucleus. Or so *classical* physics predicted. However, two physicists, Cockcroft and Walton, way back around 1930, following up a clue provided by natural radioactivity, fired particles with energies far too small to surmount these potential barriers, and found that somehow or other a fraction of the particles were able to get *through* the barriers without having to climb over the top of them – a result as I have said, totally inexplicable in terms of classical physics, and in a sense, absurd. The explanation came in terms of so-called *wave mechanics,* which postulates that the behaviour of atomic particles in certain types of interaction can be less like colliding billiard balls and much more like trains of waves. And wave mechanics was able to predict that despite the potential barrier, apparently insurmountable according to classical physics, there was a certain finite probability of bombarding particles being found *inside* the barrier. In other words, there were routes *through* the barrier which avoided climbing over it.

Is it not possible that the problem of modern man can be seen in analogous terms – that, willy-nilly, his development has been such as to create between himself and reality an intellectual barrier which, from inside it appears to be totally insurmountable, tending to drive him first towards existential despair and, beyond that perhaps, to regard life itself as a total absurdity? Yet could he but have the faith – as Cockcroft and Walton had in the *scientific* context – to attempt the seemingly absurd breakthrough, might he not find that insight and understanding, and an awareness of the subsistent reality, can enter the deeper levels of his being without having to surmount the intellectual barrier which at present looms disproportionately large for him?

Indeed, before moving on, we might remind ourselves that despite the layman's impression, even in science cold reason by no means holds total sway. As Sir Peter Medawar pointed out a few years ago in a broadcast, the average scientific paper is what he called a 'sort of fraud', which formed part of an unthinking conspiracy among scientists to misrepresent the nature of the scientific process to the world at large, a process which, according to John Davy of 'The Guardian', when reviewing 'The Double Helix' by James Watson, is portrayed by implication as 'the inexorable progress of an impersonal steamroller, steadily advancing to flatten the confusion of nature into a firm tarmac of established laws'. 'In the process', said Davy, 'human nature is flattened too: there is no scope, beneath these majestic rollers, for imagination or prejudice, for passion or speculation, for personality or that strange alchemy whereby two minds are often better than one'. Davy goes on to say that Watson's book illustrates not only the importance of the *personality* of the scientist as such, but also brings out 'a central truth about science itself, *as actually practised*, namely that 'it always starts as an *imaginative* exploit – the *apprehension* of what *might* be, rather than the observation of what *is* '. It doesn't represent much of a shift in terminology, does it, to speak of 'the *faith* that is the substance of things hoped for?'

To spend a moment or two now in further recapitulation, before passing on to the final phase of this talk.

So far we have reminded ourselves that the unconscious, the non-rational part of ourselves, has great depths, normally unplumbed by our conscious selves, depths which have incalculable negative and positive influences on our conscious selves, our choices and actions, whether we be scientist or layman, depths which we are all to a greater or lesser extent apprehensive of, and against the possibility of being over-run by which we all of us operate a variety of defences – whose particular *modus operandi* we choose subconsciously, so that we feel most at home within our conscious selves, and our immediate 'milieu'. The latter will include our homes and families, neighbours, friends, colleagues and work, the choice of some or all of which will be governed by the kind of psychological support which best meets our needs. This much is barely disputable.

However, despite the fact that the rationalist probably thinks of his philosophical position vis-à-vis that of his contestants as an *attacking* one, it has been my contention that the rationalist position is in the last analysis itself just one among the many *defensive* positions which men adopt. To the rationalist, modern man's basic difficulties – particularly those involved in a religious view of life – are seen as intellectual ones, and religion in consequence is usually given short shrift, if not actually summarily dismissed with contempt. Even Alex Comfort (who is by no means a rationalist in the ordinary sense), in his book of over 200 pages, and with the title 'Nature and Human Nature', devotes *in toto* less than two pages to Christianity *per se*, with further scattered references of a few lines apiece; and the *word* 'Christianity' does not even feature in the Index. Little wonder, in a way, since his notion of Christianity seems to be of a distinctly Victorian brand – of the 'pie in the sky' variety. For example, he defines religious *belief* - as distinct from religious *experience* of the aboriginal 'dream-time' variety - as 'an act of self-indulgence, a willingness to settle for a *lesser* intellectual integrity in order to escape from the cold of insecurity, indeterminacy, and unfulfilment'. So that's what happened to Bonheoffer, Niemüller, Schweitzer, Trevor Huddleston, Tielhard de Chardin and Martin Luther King, to mention but a handful! But I disagree. My point, really, is that to see modern man's dilemma as a purely *intellectual* one is an example, *par excellence,* of 'cutting one's facts to suit one's emotional limitations', and that it may, in fact, represent a subconscious desire to close one's eyes to the *existential* aspects of the dilemma, the attempt to deal with which may involve getting out of one's intellectual armchair, taking off one's slippers, putting on one's spiritual gum-boots, and venturing forth into the muddy plains below. Little use would it be there, to hand out stony flints hewn from the battlements of one's Ivory Tower of Pure Intellectualism, when even a few crumbs of the bread of life would spell salvation from existential anxiety and incipient despair. It was during the course of Dr Chapman's lucid and extremely helpful lecture on 'New Theology and the Search for Reality', given here before Christmas, that I first saw this possible basic difference between the reactions and approach of at least *some* of the more intellectually-orientated of the 'Death of God' theologians on the one hand, and the existential theologians typified by Paul Tillich on the other. As Dr Chapman said, the approach of the former is 'objective and unemotional – and consequently gives the impression of having less depth, and of being over-intellectualized'. The existential theologians, on the other hand, see the basic problem of modern man as one of *estrangement* – estrangement from others and estrangement from his own inner, non-rational depths, an estrangement which I believe personally has to a large extent arisen from the over-emphasis – and I almost said 'over-rating' - of the rational at the expense of the non-rational, so that communication between the bridge and the engine-room have become dangerously attenuated in most of us; an estrangement which, to use Jung's phrase again, is already taking 'a bitter revenge', leading to despair, and beyond that to the point where, finally, life is felt to be a total absurdity.

In this context we must be careful not to fall into the trap of thinking of such states as necessarily *pathological*, pointing at least to neurosis if not psychosis, and requiring the attentions of a physician or psychiatrist rather than of a theologian or a priest. With Frank Lake, I think otherwise. He says: 'It is only by *firm disassociation of thought* about our end that we can remain oblivious to the *realistic* anxiety it engenders, and *must* do, *in any reflective person*' (italics mine). "In some elderly folk', he goes on to say, 'and some not at all old, the *existential* anxieties which call in question our final state, ultimate being and meaning, and the verdict of the last judgement on our guilty lives, may be so pressing (even though not a word about them is spoken to anyone – the "morbid", we know, are shunned) that a primal anxiety state resonates through into consciousness'. Cutting deeper still, Paul Tillich says, 'Sin in its most profound state, sin as despair, abounds among us'.

This kind of despair is portrayed vividly in the writings of such men as Camus and Sartre. The latter's despair seems to have reached the end of the road where he says 'Hell is other people', whilst to Camus, according to Adele King, the world seemed 'unbearably cruel because its beauty is temporary and because its promise of happiness is cut short by death'. Yet despite such feelings Camus at least *wanted* men to continue to fight for happiness, and to attempt to preserve their dignity by balancing a state of permanent rebellion against the world's injustice with a consent to its existing beauty. But despite this, 'Conquest or play-acting, multiple loves, absurd revolt', he says, 'are tributes that man pays to his dignity in a campaign in which he is defeated in advance'.

Such despair and sense of ultimate futility and absurdity is by no means confined to the world of literature. It is implicit in the feverish and fretful quest for distraction which characterizes so much of modern life – and never before has such distraction been so readily available – literally at the turn of a knob, and in glorious technicolour, too, on all three channels.

When all argument has abated, and all debate has died, to whom then *shall* we turn for salvation from the despair and futility which abounds in the face of the one certainty which emerges unchallenged from it all – that, as Alex Comfort puts it, 'when it comes to the crunch, we personally will die?' And he adds: 'It is no answer to the dislike I feel for the idea of dying that my *genes* will persist to produce higher achievements, or that my thoughts will join a kind or moral Van Allen belt around the human race' …. 'Death', he goes on, 'ours and others', means separation, and we are social animals for whom "parting is all we know of heaven, and all we need of hell". 'This is', he says, 'like so many of our deepest feelings, an *irrationality* from the standpoint of Pure Mind – but it is also part of the predicament which makes us human …. There are times when we ought to weep – and there are times when a frank admission of loneliness or fear is more "human" than any amount of inspiring Pollyanna'.

To whom then *shall* we turn if we are not to be caught up in the Pollyanna? Freya Stark, in an essay on Education, asks the simple but profound question: 'Who dares to be intellectual in the presence of death?' Perhaps in posing the question she has at the same time implied, if not the answer, then at any rate the direction in which it would be fruitless to look for one.

Dr Chapman quoted Van Buren as saying: 'The language of existentialist theologians seems strange to the man whose job, community and daily life are not in the context of the pragmatic, empirical thinking of industry and science, *except perhaps in moments of exceptional personal crisis*' (italics mine). This qualification is vital, for it is surely the 'moments of exceptional personal crisis' which are crucial for us all. I do not believe that the ultimate answers – if there are any – to the issues raised by the problems of life and death and the anxieties they generate, will ever be formulated satisfactorily in purely intellectual terms. In fact, I believe that the *attempt* to formulate the answers in these terms is quite largely responsible for the growing existential despair among us. It was said in one of the Guildford

Cathedral lectures last autumn that there is a real danger that we are being left behind emotionally and spiritually in a maelstrom of technological innovation. If we are to find peace, there must be a real reconciliation of the conscious, rational selves *responsible* for the innovations, and the unconscious, non-rational depths which have *motivated* them. At present, our 'busy-ness' on the conscious level prevents most of us from even having the opportunity to pass the time of day with his deeper non-rational self; and we know little or nothing of its real needs, or even the ABC of its language or the rudiments of its own, special logic. Such a reconciliation would, I believe, profoundly modify our priorities, and so change the direction of even our technological drives, that our world would soon become a much more truly human and humane place to live in than it is at present. But to achieve this reconciliation we need the power to say the vital 'yes' to our deeper selves of which Paul Tillich speaks: 'Then peace enters into us, and makes us whole, self-hate and self-contempt disappear, and self is re-united with itself. ...... Then we can indeed say that "grace has come upon us".'

But the acceptance of grace as a gift from God requires one thing of us – humility. Indeed, before we can even stretch out our hand to receive the gift, humility will be needed to recognize our *need* of it. Somewhere in his book, Alex Comfort says in another context that 'to the discipline of integrity required by science we shall have to add a discipline of humility', and he continues: 'This should come pretty naturally from our subject, if we are learning much from it ourselves.' I am afraid I do not entirely share this optimism. Whilst humility is certainly an essential part of the make-up of any truly *great* scientist, it is neither an essential nor a particularly conspicuous characteristic of the technological mind – a fact which I believe is to our detriment and potential danger. But if and when humility ever finally forsakes the *pure* scientist – then it will indeed be time to flee to the mountains.

A SCIENTIST'S APPROACH TO THE RESURRECTION

It *could* be a scientist speaking:
'Except I shall see in his hands the print of the nails and put my finger into the print of the nails and thrust my hand into his side I will not believe'.

Jesus replied: "Be not faithless but believing. Blessed, [surely he meant in some special way] are they that have not seen and yet have believed'.

Is it stretching the story of Thomas overmuch to regard it as a parable of the scientist who considers that he must have a scientific explanation (whatever we mean by that) of the resurrection as a pre-condition to his embarking on the experiment of a life of faith? We might then take Christ's words: 'Blessed are they that have not seen' as 'blessed are they that have not been *able* to understand or *demanded* to understand in the scientific manner and yet have believed': that is, have taken the leap into the darkness which inevitably shrouds the beginning of the venture of faith.

In Thomas Merton's words 'It is in this darkness that we see the insufficiency of our greatest strength and in which we have nothing of our own to rely upon …. and nothing in the world to guide us or give us light … it is in this darkness, when there is nothing left in us that can please or comfort our *minds* …. that we find true liberty. It is in this abandonment that we are made strong …. this is the night that empties us …. and when we are stripped of the riches which were not ours …. *and when we rest even from that good and licit activity of knowing and desiring* which still could not give us any possession of our true end and happiness, *then* we become aware that the whole meaning of life is a poverty and emptiness which, far from being a defeat, are really the pledge of all the great supernatural gifts of which they are the potency. We become like vessels that have been emptied of water that they may be filled with wine. …. Once we begin to find this emptiness, no poverty is poor enough …. *the more our faculties are emptied of their desire and tension towards created things*, the more they collect themselves into peace and interior silence and reach into the darkness where God is present to their deepest hunger …..'

So much for Thomas Merton. The darkness which enshrouds a mystery is no new experience to the scientist, who daily lives with it in his laboratory, where it generates a creative tension within him which fertilizes his imagination and arouses his intuition. And the journey into the darkness of which the mystics speak is certainly no threat to the integrity of the scientist. Indeed, the willingness to undergo it is, I believe, peculiarly necessary to the scientist who is overburdened with the scientific approach, and in undergoing it he will become a *more*, not *less*, whole person.

I say this in spite of the widespread belief that our problem as far as the presentation of our Christian faith is concerned is basically an intellectual one, by which I presume is meant that we are up against the problem of propounding our faith in terms of pure reasoning. I hope I have already said enough to indicate the nature of my reasons for believing that this is not possible, and never will be, and that the attempt to do it not only tends to be self-vitiating, but actually destructive of the kind of knowledge we are endeavouring to impart.

Let me speak a little more personally, in a great spiritual crisis in my life, many years ago, I was given, in a deeply mystical experience, these words to ponder: 'Be still and know that I am God.' I believe now that this is a text of special value and significance to anyone who is preoccupied with the scientific approach to life.

But to move on: In that same spiritual crisis it was also given to me as in a blinding flash (which, as George Fox would have put it, 'spoke to my condition') that it was not *life* and death which were opposites and mutually exclusive, but *love* and death; that our deepest experience of love − of agapé, and all that is true and lovely and of good report in Eros − and of beauty, and truth, and goodness, and holiness, which are so intertwined with our experience of love, *all* these are at the very heart and meaning of existence itself; that the choice before us is a choice, on the one hand between death as the end of all, and on the other, of yielding ourselves to the evidence provided, not by our five senses, but by the quality of the response of the deepest level of our beings to these, the deepest experiences of our lives. Once we begin to respond at this level we begin to know that anything which superficially appears to make an end of love is itself a fraud and a usurper. And of course it is *death* which plays just this role in the minds of countless millions. Love as we know it between husband and wife, parents and children, and in the richness of friendship, is just so much poignant nonsense if death *is* the end of it.

Now it is the sheer incongruity between death on the one hand and our deepest experience of love on the other that faces us all, ultimately, scientist and non-scientist alike, with the choice between living in the faith that the totality of our lives (which includes our life in the laboratory and our life in the home) is a *meaningful pattern, and whole,* and living in the last analysis *pointlessly,* believing that it is all 'sound and fury, signifying nothing'. That is the stark choice, and I am sure that it is a crucial one for scientists, to whom a sense of coherence — or call it what you will — means so much.

Now, and most particularly, it is just the resurrection of Christ that makes sense out of what, otherwise, would be the ultimate nonsense of our deepest experiences of human life. And with our minds travelling along the lines I have indicated, *it is for evidence in human history of such a resurrection, following a unique act of love, that we would look, in order to make sense out of the apparent nonsense.*

We *experience* love, and goodness, and beauty, and holiness; we neither need nor ask for scientific *explanations* of them. A husband doesn't demand of his wife a scientific explanation for her intense emotional fixation on him, nor does one demand such an explanation of the sense of sublimity one may have listening to Beethoven's Ninth Symphony. We *experience* these things and rejoice in them, and we know if we search our hearts as well as our heads that to experience them is what matters, and that any attempt which we might foolishly embark upon at *explaining* them in scientific terms would prove both fruitless and irrelevant. Put another way: experienced at first hand, love, and those other qualities I have referred to, are completely self-substantiating and self-validating, and definitive in our lives.

And the victory of the resurrection is the victory, not just of *life*, but of transforming *love*, operating for the first time in a human life completely unimpeded by self-interest, in a demonstration, unique in history, that love is stronger than death, overcoming the last enemy.

There is much more that I would say, but I must confine myself to just one more aspect of the matter, which again is crucial to me. I once came across this sentence: 'The amount of science of which an individual scientist is ignorant is only slightly less than that of which the *non*-scientist is ignorant'. Now I believe that this simple fact has profound implications. I am a physicist, but it is only a very small area of physics in which I could claim to have real expertise, and most of us scientists would have to make a similar admission in relationship to our respective disciplines. If this is true within a single discipline, what of the fundamentally different disciplines? Which discipline then should we address in presenting a scientific explanation of the resurrection as and when we had produced one, and which discipline could we hope to satisfy with our explanation? Would it be the biologists, or biochemists, the biophysicists, the medicals, the psychologists – or possibly even the cosmologists and mathematicians? If the reply is that of course it could not be to any *one* of these disciplines that we would address ourselves exclusively, but to a sort of committee of experts, and that the specialists comprising such a body would naturally accept the verdicts of their co-specialists in the other disciplines – if *that* is our reply then we have arrived at the position where we have admitted that whatever progress we make with our scientific 'explanation' we have recognized at the outset that we cannot satisfy *a single individual scientist at first hand, within his own person*, because, inevitably, he will have had to have accepted, on the basis of the expertise of others, whole areas of our 'explanation'.

Now, before you get too impatient with me, let me assure you that I am not just playing with words or stating the obvious; for me this is at the heart of the matter. I do not believe that the life of the spirit *can* be lived at *second hand*, vicariously, *or* underwritten by the expertise of others, scientific, theological or otherwise. Each of us must make his own personal journey, each make his own mistakes, each learn from them; and above all, and in particular, in Thomas Merton's words, there *is* a sense in which we must be emptied before we can be filled.

And how *specially* true this is for the scientist!

APPENDIX 4

KEMP, Lloyd Asquith Winston Ewart

Recommendation for promotion on special merit

1. National Physical Laboratory, Teddington

2. BSc 1st class Honours (London) 1935, Postgraduate Diploma in Education (London) 1936, Ph.D. (London) 1951. F.Inst.P.

3. Temporary P.S.O.

4. S.P.S.O. (Special Merit)

5. 1932-36   King's College, London
   1936-39   Research Physicist, G.E. Research Laboratories, Wembley
   1939-41   Lecturer, The Polytechnic, Regent Street, W.1.
   1941-44   Physics Master, Bradford Grammar School
   1944-66   Physicist, The London Hospital
   1966 -   Temporary P.S.O., N.P.L.

6. Radiation dosimetry

7. Dr. Kemp is in charge of a group consisting of 1 S.C., 2 H.E's and 2 S.A.'s. He reports to Dr. W.A. Jennings, as organisational S.P.S.O. No change would be made in the organisational structure if this application for special merit promotion is successful.

9. Dr Kemp has a well deserved and formidable reputation in the field of scientific instrumentation due, in part at least, to his characteristic approach to problems by starting from first principles rather than accepting currently held notions. A classic

example of this resulted in his pointing out, in 1953 (Ref. 14 and 15), that most major national laboratories, including the NBS and, embarrassingly, the NPL, were in serious error in the measurement of the röntgen. Much water has of course passed under the bridge since then, but for someone who has the ability to uncover an international *faux pas* of such subtlety there is much poetic justice in the present recommendation.

Kemp has long been associated with new developments in the field-guarding techniques and was the first to introduce this concept in the design of cavity ionisation chambers when he introduced his "peg-less" type of cavity chamber in 1956 (Ref.18). Since the electric field within such a chamber is much more uniform than in the conventional chamber the saturation characteristics are amenable to mathematical analysis, thus removing a source of uncertainty in the measurement of dose particularly when using pulsed radiation as delivered by a linear accelerator.

On joining the NPL Kemp took charge of a group concerned with X-ray dosimetry, and he embarked on critical appraisal of the existing free-air chamber standards and the correction factors associated with them; these results were summarised in a series of internal reports. Among the problems which emerged was an apparent anomaly between the NPL and the NBS field distortion correction factors which he subsequently demonstrated to be due to compensatory effects on field distortion caused by the end-plates and side walls of the casing, (Ref.31).

The NPL is committed to develop primary and secondary standards at 'protection' exposure levels, and Kemp has turned his attention to both these problems. For the former, he has initiated basic research into the properties of the hitherto neglected cylindrical, as against parallel plate, free-air ionisation chambers. Such research includes analogue techniques for the investigation of the electrostatic field distribution, and initiating a computer programme for the analysis of the ionisation losses entailed. In addition to certain physical advantages related to cylindrical geometry, construction of the chambers should be much simpler. Kemp's approach to the problem of devising a secondary standard at 'protection' exposure levels is particularly novel and again demonstrates the attitude mentioned earlier. Recalling his former hospital experience, he appreciated the suitability of the re-breathing bag attached to anaesthetic equipment to act as an ionisation chamber with particularly appropriate characteristics (Ref.34). Such a bag has 'built-in' conductivity in its wall material, and is relatively robust and cheap. He has shown that an ionisation chamber formed by blowing up such a bag to a mere 1% above atmospheric pressure has both excellent energy and directional characteristics, and gave virtually 'free-air' performance. The relatively large volume gives high sensitivity and thus straightforward ancillary electronics will suffice. Much interest has been expressed regarding the potential applications of this instrument, not only as a secondary standard, but for field use in health physics measurements generally.

Measurements at radiotherapeutic levels have also attracted Kemp's attention; again he has designed a dosimeter specifically conceived as a secondary standard to form an essential link in the dissemination of standards from the NPL to the user. This dosimeter incorporates a number of improvements and features which are not found in a field instrument. The Ministry of Health has been so impressed by the advantages of this development that it has already agreed to place an order for 30 such instruments to cover all the NBS designated centres, providing field trials prove successful. In addition there is likely to be a small but significant export market for this instrument when it becomes available commercially in about a year's time.

For many years, Kemp has been known for the development of a comparator technique for measuring the ratio of two currents, a technique which he first introduced in 1946, indeed the literature since then contains numerous references to the use of 'Kemp Comparators'. The instrument enables the ratio of two currents, for example from a measuring and a monitoring chamber, to be determined much more rapidly and usually with greater precision than measurements on the individual currents would allow. It was clear that such an instrument would be singularly appropriate for calibration purposes at NPL, for the basic need is to compare a standard chamber to an unknown one via a monitor chamber. To this end, Kemp constructed a precision version of his brain-child, incorporating a new balanced circuit for use with electrometer tetrodes. In addition, he made a study of a special type of high insulation dry-reed relays for use as an earthing key, and their use has enabled quite complex ganged switching under electrometer conditions. The complete instrument has now been built, full consideration being given to the ergonomic aspects of the console. The latter includes not only the comparator, but also digital calibration, display and print-out facilities. As an example of its usefulness the measurement of X-ray absorption curves in aluminium and copper may be cited. The measurements were completed in three days using the comparator, but would have taken at least ten times as long had conventional techniques been employed and even then the overall precision achieved would have been markedly less.

Outline of future work and an appreciation of its significance

In the past the physical parameter chosen to quantify a radiation field has been the röntgen, a unit which is based on the ionisation produced by electromagnetic radiation in air. A considerable amount of radiotherapeutic expertise has been accumulated in terms of this unit, but it becomes increasingly difficult to measure in terms of this unit as the photon energy is increased; there are, moreover, cogent reasons for using such high-energy radiations in radiotherapy. For this reason the International Commission on Radiation Units and Measurements has introduced a new unit - the rad - which is based on the energy deposited in a given mass of material. This has obvious advantages; it is not restricted to air, and is applicable to all types of ionising radiations. The rad however is extremely difficult to establish in absolute terms since the amount of energy one is considering is very small. For example, the amount of radiation which is lethal to a human being would raise the body temperature by only a few thousandths of a degree Centigrade. It is further significant that no national laboratory has yet established a rad standard. However Kemp has shown great interest in this problem and has already conceived a design for a microcalorimeter which includes new and novel techniques. Undoubtedly his future lies in developing the highly specialised instrumentation which the change over to the rad will entail.